Methods for Unconstrained Optimization Problems

Modern Analytic *and* Computational Methods *in* Science *and* Mathematics

A GROUP OF MONOGRAPHS
AND ADVANCED TEXTBOOKS

Richard Bellman, EDITOR
University of Southern California

Published

1. R. E. Bellman, R. E. Kalaba, and Marcia C. Prestrud, Invariant Imbedding and Radiative Transfer in Slabs of Finite Thickness, 1963

2. R. E. Bellman, Harriet H. Kagiwada, R. E. Kalaba, and Marcia C. Prestrud, Invariant Imbedding and Time-Dependent Transport Processes, 1964

3. R. E. Bellman and R. E. Kalaba, Quasilinearization and Nonlinear Boundary-Value Problems, 1965

4. R. E. Bellman, R. E. Kalaba, and Jo Ann Lockett, Numerical Inversion of the Laplace Transform: Applications to Biology, Economics, Engineering, and Physics, 1966

5. S. G. Mikhlin and K. L. Smolitskiy, Approximate Methods for Solution of Differential and Integral Equations, 1967

6. R. N. Adams and E. D. Denman, Wave Propagation and Turbulent Media, 1966

7. R. L. Stratonovich, Conditional Markov Processes and Their Application to the Theory of Optimal Control, 1968

8. A. G. Ivakhnenko and V. G. Lapa, Cybernetics and Forecasting Techniques, 1967

9. G. A. Chebotarev, Analytical and Numerical Methods of Celestial Mechanics, 1967

10. S. F. Feshchenko, N. I. Shkil', and L. D. Nikolenko, Asymptopic Methods in the Theory of Linear Differential Equations, 1967

11. A. G. Butkovskiy, Distributed Control Systems, 1969

12. R. E. Larson, State Increment Dynamic Programming, 1968

13. J. Kowalik and M. R. Osborne, Methods for Unconstrained Optimization Problems, 1968

14. S. J. Yakowitz, Mathematics of Adaptive Control Processes, 1969

15. S. K. Srinivasan, Stochastic Theory and Cascade Processes, 1969

16. D. U. von Rosenberg, Methods for the Numerical Solution of Partial Differential Equations, 1969

17. R. B. Banerji, Theory of Problem Solving: An Approach to Artificial Intelligence, 1969

18. R. Lattès and J.-L. Lions, The Method of Quasi-Reversibility: Applications to Partial Differential Equations. Translated from the French edition and edited by Richard Bellman, 1969

19. D. G. B. Edelen, Nonlocal Variations and Local Invariance of Fields, 1969

20. J. R. Radbill and G. A. McCue, Quasilinearization and Nonlinear Problems in Fluid and Orbital Mechanics, 1970

21. W. Squire, Integration for Engineers and Scientists, 1970

22. T. Parthasarathy and T. E. S. Raghavan, Some Topics in Two-Person Games, 1971

23. T. Hacker, Flight Stability and Control, 1970

24. D. H. Jacobson and D. Q. Mayne, Differential Dynamic Programming, 1970

25. H. Mine and S. Osaki, Markovian Decision Processes, 1970

26. W. Sierpinski, 250 Problems in Elementary Number Theory, 1970

27. E. D. Denman, Coupled Modes in Plasmas, Elastic Media, and Parametric Amplifiers, 1970

28. F. H. Northover, Applied Diffraction Theory, 1971

29. G. A. Phillipson, Identification of Distributed Systems, 1971

30. D. H. Moore, Heaviside Operational Calculus: An Elementary Foundation, 1971

31. S. M. Roberts and J. S. Shipman, Two-Point Boundary Value Problems: Shooting Methods

32. V. F. Demyanov and A. M. Rubinov, Approximate Methods in Optimization Problems, 1970

33. S. K. Srinivasan and R. Vasudevan, Introduction to Random Differential Equations and Their Applications, 1971

34. C. J. Mode, Multitype Branching Processes: Theory and Applications, 1971

35. R. Tomovic and M. Vukobratovic, General Sensitivity Theory

36. J. G. Krzyz, Problems in Complex Variable Theory

37. W. T. Tutte, Introduction to the Theory of Matroids, 1971

38. B. W. Rust and W. R. Burrus, Mathematical Programming and the Numerical Solution of Linear Equations

39. J. O. Mingle, The Invariant Imbedding Theory of Nuclear Transport.

40. H. M. Lieberstein, Mathematical Physiology.

Methods for Unconstrained Optimization Problems

by

J. KOWALIK

Computing Department
The Boeing Company, Renton, Washington

AND

M. R. OSBORNE

Computer Centre
The Australian National University
Canberra, A.C.T., Australia

American Elsevier Publishing Company, Inc.

New York 1968

AMERICAN ELSEVIER PUBLISHING COMPANY, INC.
52 Vanderbilt Avenue, New York, N.Y. 10017

ELSEVIER PUBLISHING COMPANY,
335 Jan Van Galenstraat, P.O. Box 211
Amsterdam, The Netherlands

International Standard Book Number 0-444-00041-0
Library of Congress Card Number 68-24805

191,546

b|9787

CONTENTS

CHAPTER 3

Descent Methods

CHAPTER 4

Least Squares Problems

CHAPTER 5

Constrained Problems

CHAPTER 6

Numerical Results

PREFACE

Methods for *computing* the minimum of a function of several variables are important not only because they can be applied to solve many problems of considerable practical importance but also because they provide the numerical analyst with a tool for tackling a wide range of nonlinear problems. It is not surprising, then, that there has been considerable recent work devoted to the development of such methods, and the pace of this development can be gauged by comparing this book with one with a similar title (D. J. Wilde, *Optimum Seeking Methods*, Prentice-Hall, Englewood Cliffs, N.J.) published in 1964. There is staggeringly little material common to both.

There seem to be good reasons why this rate of development will continue with certain classes of methods at least, and the recently discovered transformations (see Chapter 5) which permit unconstrained techniques to be applied to solve constrained problems guarantee that the range and importance of applications will continue to grow. However, there are certain methods that have sufficiently similar derivations to permit a satisfactory unifying treatment, and this we have attempted in Chapter 3 for descent methods, and in Chapter 4 for methods for minimizing a sum of squares. These chapters include the most important of current methods so that the possibility of unifying treatments indicates that an important stage has been reached in the evolution of the subject. Chapter 2, in which we treat the extremely important direct search methods, pretends to no such unifying treatment. In our numerical experiments (summarized in Chapter 6) we have found the simplex method in particular to be surprisingly successful. It seems to us that appropriate implementation can offset to a certain extent at least the alleged decrease in efficiency of these methods as the dimensionality of the problem is increased. There seems to be room for much research and experiment here.

The presentation of material assumes that the reader has some facility with matrix algebra. We summarize our notation and the principal results we require in Appendix 1. We also use some properties of convex sets (in Chapter 5 in particular) and these are summarized in Appendix 2.

We should like to record our indebtedness to Mr. Alastair Watson, who has generally assisted with discussion of techniques, preparation of examples, and checking of the manuscript, to Mr. John Boothroyd, who has made a number of suggestions that have greatly improved the presentation, to Gail Liddel and Stephanie Percivall, who prepared the typescripts quickly and well and to Rachelle Yocum who prepared the illustrations. Finally, we should like to express our thanks to the Editor, Richard Bellman, who took the initiative in suggesting that the book be written.

J. KOWALIK
M. R. OSBORNE

Canberra
June 1968

Chapter I

PRELIMINARY CONSIDERATIONS

I.I. MAXIMA AND MINIMA

In this book we regard the words optimize and minimize as synonomous.

DEFINITION. We say $\mathbf{x}_0$ is a *minimum* of the function $F(\mathbf{x})$ (the objective function) if there is a region R containing $\mathbf{x}_0$ in its interior such that

$$F(\mathbf{x}) \geqslant F(\mathbf{x}_0), \qquad \mathbf{x} \in R. \tag{1.1.1}$$

If

$$F(\mathbf{x}) > F(\mathbf{x}_0), \qquad \mathbf{x} \neq \mathbf{x}_0, \qquad \mathbf{x} \in R, \tag{1.1.2}$$

we say that $\mathbf{x}_0$ is a *proper* minimum of F.

DEFINITION. We say that $\mathbf{x}_0$ is a maximum of the function $F(\mathbf{x})$ if it is a minimum of $-F(\mathbf{x})$.

If a proper minimum is unique in an appropriate sense, then we say it is a *global* minimum, otherwise it is a *local* minimum. The techniques to be developed in subsequent chapters are for finding proper minima. The distinction between local and global minima is not essential for our purposes, and we note that a global minimum of F on a set S may be a local minimum on a set S^1 where $S < S^1$. It is, however, an important distinction when the results of applying optimization techniques have to be interpreted.

I.2. CONSTRAINED AND UNCONSTRAINED PROBLEMS

The problem of minimizing the function $F(\mathbf{x})$ for which the values of the components of $\mathbf{x}$ are not restricted by side conditions or *constraints* is called an *unconstrained* optimization problem. The classic methods of the calculus (which we assume to be applicable) give the necessary conditions

$$\frac{\partial F}{\partial x_1} = \frac{\partial F}{\partial x_2} = \cdots = \frac{\partial F}{\partial x_n} = 0, \tag{1.2.1}$$

I

which must be satisfied by a stationary value. In the sense expressed by equation (1.2.1) the problem of finding a stationary value of a function is equivalent with that of solving a system of nonlinear equations. Further, this latter problem can be posed as an optimization problem. For consider the system of equations

$$f_i(\mathbf{x}) = 0, \qquad i = 1, \ldots, n. \tag{1.2.2}$$

Then a solution to these equations also minimizes the function

$$F(\mathbf{x}) = \sum_{i=1}^{n} \{f_i(\mathbf{x})\}^2. \tag{1.2.3}$$

The relationship between these problems can often be usefully exploited. For example, consider a function $G(\mathbf{x})$ having a saddle point (a stationary value which is neither a maximum nor a minimum). The optimization techniques considered in this work are not directly applicable to finding this point; however the coordinates must satisfy the system of equations

$$\frac{\partial G}{\partial x_i} = 0, \qquad i = 1, 2, \ldots, n \tag{1.2.4}$$

and they must also minimize

$$F(\mathbf{x}) = \sum_{i=1}^{n} \left\{ \frac{\partial G}{\partial x_i} \right\}^2 \tag{1.2.5}$$

This observation is useful in solving what Hadley calls the classic constrained optimization problems. These are problems of the form

$$\text{minimize } F(\mathbf{x}) \tag{1.2.6}$$

subject to the constraints (equality constraints)

$$g_i(\mathbf{x}) = 0; \qquad i = 1, 2, \ldots, m. \tag{1.2.7}$$

The classic Lagrange multiplier technique can be used to write down necessary conditions which a stationary point must satisfy, and it turns out that such a point makes stationary the unconstrained function

$$G(\mathbf{x}, \boldsymbol{\lambda}) = F(\mathbf{x}) + \sum_{i=1}^{m} \lambda_i g_i(\mathbf{x}). \tag{1.2.8}$$

Let $(\mathbf{x}_0, \boldsymbol{\lambda}_0)$ be a stationary point of $G(\mathbf{x}, \boldsymbol{\lambda})$. Then this is a saddle point, for in a neighborhood of this point we have [1, p. 75]:

$$G(\mathbf{x}_0, \boldsymbol{\lambda}) \geqslant G(\mathbf{x}_0, \boldsymbol{\lambda}_0) \geqslant G(\mathbf{x}, \lambda_0). \tag{1.2.9}$$

Thus the classic constrained problem can be approached either

(i) by solving the system of equations

$$\frac{\partial F}{\partial x_i} + \sum_{j=1}^{m} \lambda_j \frac{\partial g_j}{\partial x_i} = 0, \qquad i = 1, 2, \ldots, n, \tag{1.2.10}$$

$$g_i = 0, \qquad j = 1, 2, \ldots, m, \tag{1.2.11}$$

or
(ii) By optimizing the function

$$H(\mathbf{x}, \boldsymbol{\lambda}) = \sum_{i=1}^{n} \left\{ \frac{\partial F}{\partial x_i} + \sum_{j=1}^{m} \lambda_j \frac{\partial g_j}{\partial x_i} \right\}^2 + \sum_{j=1}^{m} g_j^2. \tag{1.2.12}$$

If the constraint conditions are relaxed to permit inequality constraints, then the optimiziation problem takes the form

$$\text{minimize } F(\mathbf{x}) \tag{1.2.13}$$

subject to the constraints

$$g_i(\mathbf{x}) \leqslant 0, \qquad i = 1, 2, \ldots, m \tag{1.2.14}$$

and (frequently)

$$x_i \geqslant 0, \qquad i = 1, 2, \ldots, n. \tag{1.2.15}$$

In this case it is called a *mathematical programming* problem. To transform this into an unconstrained optimization problem similar to equation (1.2.12) it is necessary to have a mechanism that indicates which of the constraints are active (that is, which of the constraints currently hold as equalities). The additional computational difficulties to which this gives rise are sketched in Hadley [1, p. 71]. These are such as to make the Lagrange multiplier approach unattractive in general; however it is likely that mathematical programming problems will frequently be solved by associating them with a *sequence* of unconstrained optimization problems, and this possibility is explored in Chapter 5.

Necessary conditions for $\mathbf{x}_0$ to be a stationary value of the objective function subject to inequality constraints have been given by John [2]. For the problem specified by equations (1.2.13) and (1.2.14) these are

$$\lambda_i \geqslant 0, \qquad i = 0, 1, \ldots, m, \tag{1.2.16}$$

$$\lambda_0 \frac{\partial F}{\partial x_j} + \sum_{i=1}^{m} \lambda_i \frac{\partial g_i}{\partial x_j} = 0, \qquad j = 1, 2, \ldots, n, \tag{1.2.17}$$

$$\lambda_i g_i(\mathbf{x}_0) = 0, \qquad i = 1, 2, \ldots, m. \tag{1.2.18}$$

Kuhn and Tucker [3] have given a "constraint qualification" which guarantees that $\lambda_0 > 0$ in equation (1.2.17) so that it can be replaced by 1. For our purposes (an application of the Kuhn-Tucker result is made in Chapter 5) it is sufficient that the $g_i(\mathbf{x})$ are convex (Appendix 2), and that the set of $\mathbf{x}$ such that $g_i(\mathbf{x}) < 0$, $i = 1, 2, \ldots, m$, is nonempty.

We call the system specified by equations (1.2.13), (1.2.14) the *primal* problem. To every convex primal problem we can associate a *dual* problem [4]:

$$\text{minimize } G(\mathbf{x}, \boldsymbol{\lambda}) = F(\mathbf{x}) + \sum_{i=1}^{m} \lambda_i g_i(\mathbf{x}) \qquad (1.2.19)$$

subject to the constraints

$$\frac{\partial F}{\partial x_j} + \sum_{i=1}^{m} \lambda_i \frac{\partial g_i}{\partial x_j} = 0, \qquad j = 1, 2, \ldots, n, \qquad (1.2.20)$$

$$\lambda_i \geqslant 0, \qquad i = 1, 2, \ldots, m. \qquad (1.2.21)$$

The basic results relating the optimum values of the primal and dual problems are:

(i) If $\mathbf{x}_0$ solves the primal problem then there exists a $\boldsymbol{\lambda}_0$ such that $(\mathbf{x}_0, \boldsymbol{\lambda}_0)$ solves the dual, and

(ii) $\max G(\mathbf{x}, \boldsymbol{\lambda}) = \min F(\mathbf{x})$.

1.3. LINEAR AND NONLINEAR PROBLEMS

If the objective function is a quadratic form

$$F(\mathbf{x}) = a + \mathbf{b}^T \mathbf{x} + \tfrac{1}{2} \mathbf{x}^T A \mathbf{x}, \qquad (1.3.1)$$

then the conditions for a stationary value take the form

$$\frac{\partial F}{\partial x_j} = \mathbf{e}_j^T \{\mathbf{b} + A\mathbf{x}\} = 0, \qquad j = 1, 2, \ldots, n \qquad (1.3.2)$$

If A is nonsingular there is a unique stationary value which is found by solving the set of linear equations

$$A\mathbf{x} + \mathbf{b} = 0. \qquad (1.3.3)$$

In this case the stationary value can be found in a finite number of operations, and we say that the problem is *linear*. Note that this term does not indicate linearity of the objective function, only that the solution can be found by solving a set of linear equations.

The case of a linear objective function is not of interest in the unconstrained case, as it can have a stationary value only if it is constant. Thus, excepting the quadratic case, the conditions for a stationary value lead to a set of nonlinear equations. In general the optimum will have to be found iteratively, and the problem is said to be *nonlinear*.

However, the optimization of a linear objective function subject to linear constraints is a nontrivial problem. In this case a solution can be found in a finite number of steps by, for example, the simplex method of Dantzig [5]. This is called a *linear programming problem*. If either the objective function or any of the constraints are nonlinear, then the problem is called a *nonlinear programming problem*.

1.4 NOTES ON HISTORICAL DEVELOPMENT

The existence of optimization problems is as old as mathematics, and the first systematic techniques for the solution of these problems stem from the development of the calculus and are associated with the names of Newton, Lagrange, and Cauchy (who made the first application of the method of steepest descent). However, little substantive progress was made until the middle of this century, when development was greatly accelerated by the availability of computers, and by an increasing requirement for the solution of decision problems (stimulated in part by World War II).

In recent years unconstrained problems have been attacked successfully by a number of direct search and descent methods, and an account of this work forms the major subject matter of this book. It is interesting that the major developments in this area have been made in the United Kingdom and (starting with Rosenbrock's fundamental paper in 1960) have been largely reported in *Journal of the British Computer Society.*

A major development in the solution of constrained problems was the publication by Dantizg in 1947 of his simplex method for linear programming problems. This method and its extensions have given rise to a large number of applications to such problems as the optimum allocation of resources and the optimization of engineering systems.

For nonlinear programming problems no single technique which handles the constraints explicitly has been generally successful, and recently considerable attention has been given to methods that apply unconstrained techniques to suitably transformed problems. Early attempts go back at least to Courant [6]. Another approach suggested by Carroll [7] has been systematically exploited by Fiacco and McCormick [8] and appears most promising. This work is described in Chapter 5.

1.5. SOME APPLICATION PROBLEMS

In our experience the majority of direct applications of unconstrained minimization techniques are of a type which can be broadly labeled as "curve fitting." For example, it may be that a physical theory predicts that observations of a particular process realized experimentally should be represented by an expression of the form

$$y = F(\mathbf{x}, \boldsymbol{\alpha}) \tag{1.5.1}$$

in which the components of the vector of parameters $\boldsymbol{\alpha}$ are constants characteristic of the particular experimental configuration. Often the aim of the experiment is to determine $\boldsymbol{\alpha}$, which is sought as the minimum of

$$G(\boldsymbol{\alpha}) = \sum_{i=1}^{N} W_i (y_i - F(\mathbf{x}_i, \boldsymbol{\alpha}))^2, \tag{1.5.2}$$

where W_i is a positive weight indicating the "goodness" of the i-th observation.

This application contains the classic curve-fitting problem which is linear in the sense of Section 1.3. Here $y_i = f(\mathbf{x}_i)$ are the function values and

$$F(\mathbf{x}, \boldsymbol{\alpha}) = \sum_{i=1}^{n} \alpha_i \phi_i(\mathbf{x}), \tag{1.5.3}$$

where the $\phi_i(x)$ are the fitting functions. More complicated applications leading to nonlinear problems have recently been considered. For example, Mason [9] has obtained approximate solutions to the Blasius equation

$$L(y) = \frac{d^3 y}{dx^3} + \frac{d^2 y}{dx^2} y = 0, \tag{1.5.4}$$

subject to the boundary conditions

$$y(0) = \frac{dy}{dx}(0) = 0, \qquad \frac{dy}{dx}(\xi) \to 1, \qquad \xi \to \infty, \tag{1.5.5}$$

by minimizing

$$G(\boldsymbol{\alpha}) = \sum_{i=1}^{N} \{L(P)(\xi_i)\}^2, \tag{1.5.6}$$

where

$$P = \alpha_{n+3} + \alpha_{n+2}x + \frac{\alpha_{n+1}}{\{Q(x)\}^4}, \tag{1.5.7}$$

and

$$Q(x) = 1 + \sum_{i=1}^{n} \alpha_i x^i. \tag{1.5.8}$$

It proves to be easy to fix certain of the α_i to satisfy the imposed boundary conditions, and the unconstrained minimum of G is found with respect to the remaining α_i. The sum is taken over a set of points $0 \leqslant \xi_1 < \cdots < \xi_N < \infty$, and in practice $\xi_N < 10$ is certainly sufficient. By this means very attractive closed form approximations have been found.

A typical formulation of a constrained optimization problem arises in connection with a design process in which we look for an optimal solution selected from some restricted class of designs. An optimal solution could mean that the desired design is (for example) the cheapest, most reliable, or of least weight. In addition to requiring a criterion by means of which we can evaluate and compare various designs, we have also to consider relationships between the design and behavioral variables related to a specific design. The former are physical properties such as dimensions, materials, etc. which can, in general, be freely selected. The latter describe the behavior of the design and may be such entities as stresses and deflections. Both sets of variables are interrelated by a set of equations called the governing technology.

If we limit ourselves to technologies which can be expressed by a set of algebraic equations, then while certain of the variables are fixed by the design specification it is usual that others are free within certain limits so that an optimization process is possible. In general the equations of the governing technology do not exhaust the requirements made on the design, and these are represented by additional constraints.

We denote by $\mathbf{b}$ and $\mathbf{d}$ the vectors of the behavioral and design variables, respectively. Then the governing technology is written as the set of equations

$$h_j(\mathbf{b}, \mathbf{d}) = 0, \qquad j = 1, 2, \ldots, m, \qquad (1.5.9)$$

and the further constraints are denoted by

$$c_i(\mathbf{b}, \mathbf{d}) \geqslant 0, \qquad i = 1, 2, \ldots, p. \qquad (1.5.10)$$

We define a vector $\mathbf{x}$ (of dimension n, say) of the free variables in the problem, and we say that a particular design described by $\mathbf{x}$ is a feasible solution if this vector satisfies the equations (1.5.9) and the inequalities (1.5.10). If there is a range of feasible solutions we say the problem is well posed. The problem now requires that the design criterion

$$f = f(\mathbf{b}, \mathbf{d}) = g(\mathbf{x}) \qquad (1.5.11)$$

be optimized from among the class of feasible solutions.

REFERENCES

1. G. Hadley, *Nonlinear and Dynamic Programming*, Addison-Wesley Publishing Co., Reading, Mass., 1964, p. 5.
2. F. John, Extremum Problems with Inequalities as Subsidiary Conditions, in *Studies and Essays* (Courant Anniversary Volume), Interscience Publishers, New York, 1948, p. 178.
3. H. W. Kuhn and A. W. Tucker, Nonlinear Programming, in *Proceedings of the Second Berkeley Symposium on Mathematical Statistics and Probability*, University of California Press, Berkeley, 1951, p. 481.
4. P. Wolfe, A. Duality Theory for Nonlinear Programming, *Quart. Appl. Math.*, *19*(1961), 230.
5. G. B. Dantzig, *Linear Programming and Extensions*, Princeton University Press, Princeton, N.J., 1963.
6. R. Courant, Variational Methods for the Solution of Problems of Equilibrium and Vibration, *Bull. Amer. Math. Soc.*, *49*(1943).
7. C. W. Carroll, The Created Response Surface Technique for Optimizing Nonlinear Restrained Systems, *Operations Res.*, *9*(1961), 169.
8. An extensive list of papers by A. V. Fiacco and G. P. McCormick on this subject is given in the references to Chapter 5.
9. J. C. Mason, D. Phil. Thesis, Oxford, 1965.

Chapter 2

DIRECT SEARCH METHODS

2.1. INTRODUCTION

Direct search methods are based on a sequential examination of trial solutions which by simple comparisons give an indication for a further searching procedure. These methods require only the ability to evaluate the function at a given point and can be used for general continuous functions. Methods of this type are useful in the early stages of optimization and can provide efficiently information about a region in which a minimum is located.

In general, they do not give a rapid rate of ultimate convergence and hence are inefficient for finding a minimum with high precision. There are, however, problems in which these features of an optimization method are not considered essential disadvantages. We may have, for example, a problem in which an objective function is a result of inaccurate observations so that a precise location of the minimum is not appropriate.

The simplest but very important optimization problem is that of finding a local minimum of a function when the values of the independent variable lie on a fixed line so that the problem reduces to finding a minimum of a function of a single variable. In the next two sections we shall discuss two types of approach to this problem: (i) the Fibonacci search, and (ii) the repeated interpolation method.

For more than one independent variable we shall show a method developed by Hooke and Jeeves. This method changes the parameters one at a time starting from an initial point, but once the full series of perturbations has been completed it takes a step along the direction joining the last and the initial point.

A considerable further improvement has been suggested by Rosenbrock. In this method lower values are sought along n mutually orthogonal directions. This set of directions is then rotated so that it adapts itself to the directions of most rapid decrease of the objective function.

The methods of Hooke and Jeeves and of Rosenbrock are presented in Sections 4 and 5.

In Section 6 we shall describe a method that depends on the comparisons of function values at the vertices of a simplex, which is then modified so that it moves toward an optimum and ultimately shrinks to this point.

2.2. MINIMIZING A FUNCTION ALONG A LINE

The problem of finding a desired minimum of a function of a single variable is one of the most important of unconstrained optimization problems because this operation is basic to many of the most important techniques of more general application.

DEFINITION.   A function $f(x)$ is called *unimodal* in $a \leqslant x \leqslant b$ if it has only one stationary value (either a maximum or a minimum) in (a, b). We will always assume that the stationary value is a minimum.

In this section we consider methods for finding the stationary value of a given unimodal function which make use of function values only, and which assume only the continuity of f. The following result is basic to the development of this kind of method.

LEMMA 2.2.1.   *Let $f(x)$ be unimodal in (a, b), then it is necessary to evaluate the function at a minimum of two interior points before the stationary value can be located in a subinterval of (a, b).*

PROOF.   If the function is evaluated at the interior point x_1, then the minimum of f can lie to either side of the point. To determine this another function evaluation is necessary (see Figure 2.2.1). Let $a < x_1 < x_2 < b$; then if $f(x_1) > f(x_2)$ the minimum lies in (x_1, b), otherwise in (a, x_2).

It is clear that by repeated evaluation of f the minimum can be located to any required precision, and that the efficiency of this process depends on the choice of the points at which the function is evaluated. We note that, as one of the points used in reducing the size of the current interval must always lie in the interior of the reduced interval, it can serve as one of the trial points for the next stage. Thus, although two function evaluations are required initially, only one further function value is required for each subsequent step.

To determine an algorithm of this kind we ask: is it possible to arrange that the size of the interval containing the minimum be decreased by a constant factor τ at each step? Let the current interval be $(a^{(i)}, b^{(i)})$, and let the points at which the function is evaluated by $x_1^{(i)}$ and $x_2^{(i)}$ where $x_1^{(i)} < x_2^{(i)}$.

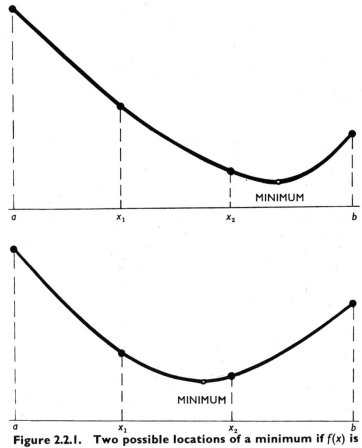

Figure 2.2.1. Two possible locations of a minimum if $f(x)$ is unimodal on (a, b) and $f(x_1) > f(x_2)$.

Then these points must satisfy

$$\frac{x_2^{(i)} - a^{(i)}}{b^{(i)} - a^{(i)}} = \frac{b^{(i)} - x_1^{(i)}}{b^{(i)} - a^{(i)}} = \tau, \qquad (2.2.1)$$

from which it follows that

$$x_1^{(i)} - a^{(i)} = b^{(i)} - x_2^{(i)}. \qquad (2.2.2)$$

If we assume that $f(x_2^{(i)}) > f(x_1^{(i)})$ then we have

$$b^{(i+1)} = x_2^{(i)},$$

and
$$a^{(i+1)} = a^{(i)},$$
and we take
$$x_2^{(i+1)} = x_1^{(i)}, \tag{2.2.3}$$
as the alternative choice is not compatible with a value of τ in equation (2.2.1) that is independent of i. This is sufficient to specify the procedure precisely, for we have

$$\frac{x_2^{(i+1)} - a^{(i)}}{x_2^{(i)} - a^{(i)}} = \frac{x_1^{(i)} - a^{(i)}}{x_2^{(i)} - a^{(i)}} = \frac{x_2^{(i)} - a^{(i)}}{b^{(i)} - a^{(i)}} = \tau, \tag{2.2.4}$$

and by equation (2.2.2)

$$x_1^{(i)} - a^{(i)} = b^{(i)} - a^{(i)} - (x_2^{(i)} - a^{(i)}), \tag{2.2.5}$$

so that

$$\frac{x_1^{(i)} - a^{(i)}}{x_2^{(i)} - a^{(i)}} = -1 + \frac{1}{\tau}, \qquad \text{by equation (2.2.5),}$$

$$= \tau, \qquad \text{by equation (2.2.4),}$$

whence

$$\tau^2 + \tau - 1 = 0. \tag{2.2.6}$$

Only the positive root of this equation is relevant, and this gives

$$\tau = \frac{\sqrt{5} - 1}{2} \approx 0.618. \tag{2.2.7}$$

The implementation of this algorithm is particularly simple. We have either

(i) if $f(x_2^{(i)}) > f(x_1^{(i)})$,

$$b^{(i+1)} = x_2^{(i)},$$

$$x_2^{(i+1)} = x^{(i)}, \tag{2.2.8a}$$

$$x_1^{(i+1)} = a^{(i)} + (1 - \tau)(b^{(i+1)} - a^{(i)}),$$

or

(ii) if $f(x_2^{(i)}) \leqslant f(x_1^{(i)})$,

$$a^{(i+1)} = x_1^{(i)},$$

$$x_1^{(i+1)} = x_2^{(i)}, \tag{2.2.8b}$$

$$x_2^{(i+1)} = b^{(i)} - (1 - \tau)(b^{(i)} - a^{(i+1)}).$$

It will be recognized that we are taking at each step the *golden section* of the current interval, and it is from this observation that this algorithm takes its name. After n function evaluations we have

$$b^{(n)} - a^{(n)} = (b^{(1)} - a^{(1)})\tau^{n-1}. \tag{2.2.9}$$

We turn now to the question of determining what algorithm of this kind (if any) is optimum in the sense that it gives the largest ratio of initial to final interval for a fixed number of function evaluations. It is clearly sufficient to assume a unit final interval, and in this case the optimum strategy is the one for which the initial interval is greatest. This question has been discussed by several authors—for example Kiefer [1] and Johnson [2].

Let L_n be the length of the largest interval that can be reduced to the unit interval in n function evaluations. It is easy to find a bound for L_n, for consider the first two function evaluations made in searching for the minimum. Let the corresponding points be x_1 and x_2 where $x_1 < x_2$. Then if the minimum lies in (a, x_1) we have at most $n - 2$ function evaluations available to refine this interval so that $x_1 - a \leqslant L_{n-2}$. However, if the minimum lies in (x_1, b) then there remain $n - 1$ function evaluations because x_2 lies in this interval so that $b - x_1 \leqslant L_{n-1}$. Thus

$$L_n \leqslant L_{n-1} + L_{n-2}. \tag{2.2.10}$$

It is clear that $L_0 = L_1 = 1$ because, by Lemma 2.2.1, no reduction of the initial interval is possible with less than 2 function evaluations. This means that the solution to the recurrence

$$F_n = F_{n-1} + F_{n-2},$$
$$F_0 = F_1 = 1, \tag{2.2.11}$$

is optimum provided only it is realizable. The F_i are the *Fibonacci numbers*. To derive an explicit form for them we seek a solution of the recurrence of the form

$$F_i = Ar_1^i + Br_2^i. \tag{2.2.12}$$

Substituting this into equation (2.2.11) we find that r_1 and r_2 must be roots of the equation

$$r^2 - r - 1 = 0, \tag{2.2.13}$$

and this gives

$$r_1 = \frac{1 + \sqrt{5}}{2}, \qquad r_2 = \frac{1 - \sqrt{5}}{2}. \tag{2.2.14}$$

The values of A and B are found by satisfying the initial conditions, and it is readily verified that the resulting form for F_i is

$$F_i = \frac{1}{\sqrt{5}} \left\{ \left(\frac{1 + \sqrt{5}}{2} \right)^{i+1} - \left(\frac{1 - \sqrt{5}}{2} \right)^{i+1} \right\}. \qquad (2.2.15)$$

Now let N be the preset number of function evaluations. Then an algorithm which realizes this optimum search is

$$x_1^{(i)} = \frac{F_{N-i-1}}{F_{N-i+1}} (b^{(i)} - a^{(i)}) + a^{(i)} \qquad (2.2.16)$$

$$x_2^{(i)} = \frac{F_{N-i}}{F_{N-i+1}} (b^{(i)} - a^{(i)}) + a^{(i)}, \qquad (2.2.17)$$

where $i = 1, 2, \ldots, N$. We have

$$x_2^{(i)} - a^{(i)} = b^{(i)} - x_1^{(i)} = \frac{F_{N-i}}{F_{N-i+1}} (b^{(i)} - a^{(i)}),$$

so that after N evaluations the length of the final interval is

$$\delta = \prod_{i=1}^{N} \frac{F_{N-i}}{F_{N-i+1}} (b^{(1)} - a^{(1)})$$

$$= \frac{b^{(1)} - a^{(1)}}{F_N}, \qquad (2.2.18)$$

and this proves that the algorithm gives the optimum interval reduction. It remains to be shown that function evaluations are economized. Let us assume that $f(x_1^{(i)}) < f(x_2^{(i)})$. Then we must show that $x_1^{(i)} = x_2^{(i+1)}$, and this follows from

$$x_2^{(i+1)} = \frac{F_{N-i-1}}{F_{N-i}} (x_2^{(i)} - a^{(i)}) + a^{(i)}$$

$$= \frac{F_{N-i-1}}{F_{N-i}} \left(\frac{F_{N-i}}{F_{N-i+1}} (b^{(i)} - a^{(i)}) + a^{(i)} - a^{(i)} \right) + a^{(i)}$$

$$= \frac{F_{N-i-1}}{F_{N-i+1}} (b^{(i)} - a^{(i)}) + a^{(i)}$$

$$= x_1^{(i)}. \qquad (2.2.19)$$

It is interesting to compare the reduction obtained by the golden section algorithm with that given by the algorithm based on Fibonacci numbers. First we note that as $|r_1| > |r_2|$ then

$$\lim_{N \to \infty} \frac{F_{N-1}}{F_N} = \frac{1}{r_1} = \tau, \qquad (2.2.20)$$

so that asymptotically the two algorithms give the same effective rate of decrease. Now let $G_i = 1/\tau^{i-1} = r_1^{i-1}$. Then G_i satisfies the recurrence

$$G_{i+1} = G_i + G_{i-1}, \qquad (2.2.21)$$

subject to the initial conditions

$$G_0 = \tau, \qquad G_1 = 1, \qquad (2.2.22)$$

and from this it follows that, for $N > 1$,

$$G_N < F_N < G_{N+1}. \qquad (2.2.23)$$

These results indicate that there is little to choose between the golden section algorithm and the algorithm based on Fibonacci numbers. By equation (2.2.23) the golden section algorithm can never require more than one additional function evaluation to satisfy a prescribed tolerance. As the golden section algorithm is somewhat simpler to implement, and as it does not require a knowledge of N in advance, it is likely to be the one chosen in practice.

It is possible for certain implementations of both algorithms to be numerically unstable (see for example Overholt [4] or Boothroyd [5]). Here the problem stems from the use of the three term recurrences (2.2.11) or (2.2.21) to generate decreasing sequences of numbers with the aim of avoiding the explicit mutiplications in equations (2.2.8), (2.2.16), and (2.2.17). For example, equation (2.2.8a) gives

$$x_1^{(i+1)} = a^{(i)} + (1 - \tau)(b^{(i+1)} - a^{(i)})$$

but $x_1^{(i+1)}$ could also be computed from

$$x_1^{(i+1)} = a^{(i)} + (x_2^{(i)} - x_1^{(i)}). \qquad (2.2.24)$$

This last equation follows by noting that

$$x_2^{(i)} - x_1^{(i)} = (x_2^{(i)} - a^{(i)}) - (x_1^{(i)} - a^{(i)})$$
$$= (1 - \tau)(x_2^{(i)} - a^{(i)})$$
$$= \tau^2(x_2^{(i)} - a^{(i)}). \qquad (2.2.25)$$

However, if we set

$$G_{N-i+1} = b^{(i)} - a^{(i)},$$

then the first line of equation (2.2.25) is equivalent to

$$x_2^{(i)} - x_1^{(i)} = G_{N-i} - G_{N-i-1}, \qquad (2.2.26)$$

and this shows that the recurrence (2.2.21) is being used in the reverse direction. Used in this way it is unstable because it has a second solution which is increasing rapidly (like $(-1/\tau)^j$). Small components of this second solution will be introduced by rounding errors and will eventually swamp the computation. This is illustrated in Table 2.2.1 in which the computed solution of the difference equation

$$H_i = H_{i-2} - H_{i-1}, \qquad (2.2.27)$$

subject to the initial conditions

$$H_0 = 1, \qquad H_1 = \tau, \qquad (2.2.28)$$

is compared with τ^i.

If equations (2.2.8), (2.2.16), and (2.2.17) are used, then instability is avoided as each new partition is computed as the correct proportion of the

Table 2.2.1.   Illustration of Use of the Unstable Recurrence
(from $i = 20$ the H_i alternate in sign)

I	H_i	τ^i
1	1.000 000	1.000 000
2	.618 034	.618 034
3	.381 966	.381 966
4	.236 068	.230 068
5	.145 898	.145 898
6	.090 169	.090 170
7	.055 729	.055 728
8	.034 441	.034 442
9	.021 288	.021 286
10	.013 152	.013 156
12	.005 016	.005 025
14	.001 895	.001 919
16	.000 670	.000 733
18	.000 116	.000 280
20	−.000 323	.000 107
22	−.001 084	.000 041
24	−.002 930	.000 016
26	−.007 706	.000 006
28	−.020 188	.000 002
30	−.052 859	.000 001
32	−.138 388	.000 000
34	−.362 306	.000 000
36	−.948 529	.000 000
38	−2.48328	.000 000
40	−6.50131	.000 000

current interval. Slightly more calculation is required per step, but this is small compared with the evaluation even of a very simple function, and is of no consequence in practice.

2.3. QUADRATIC INTERPOLATION FOR MINIMIZING A FUNCTION OF A SINGLE VARIABLE

The search methods described in the previous section have the advantage of being easy to implement and of having a guaranteed rate of convergence. Unfortunately, however, the rate of convergence is rather slow if the minimum is to be found to a high precision. For example, if we want the ratio of the final to initial interval to be $\leqslant 10^{-4}$ then N, the number of function evaluations, is determined as the index of the first Fibonacci number $\geqslant 10^4$. This number is $F_{20} = 10,949$ and consequently $N = 20$. Until now we have made only minimal assumptions concerning the behavior of the function to be minimized, and it is reasonable to expect that more efficient algorithms would be possible if more information about the function is available.

Perhaps the simplest assumption is that, in the neighborhood of the minimum, the function is adequately represented by a quadratic interpolation polynomial. This assumption is closely related to assumptions we make in Chapters 3 and 4, where we assume that in the neighborhood of the minimum the function to be optimized can be approximated by a quadratic form. Both in its present restricted form, and in the more general form used in subsequent chapters, this assumption has been thoroughly tested in practice and found satisfactory.

The basic algorithm for this method has two main forms:

(i) The function is evaluated at three points, a quadratic interpolation polynomial is fitted to it, and the minimum (hopefully) of this interpolant is formed. This point replaces one of the initial points, and the procedure is repeated until a suitable convergence criterion is satisfied. Precautions are necessary to prevent either divergence or convergence to a maximum and a description of a suitable algorithm is given by Powell [6].

(ii) In this approach a region containing the minimum is first bracketed by a systematic search technique. This bracket is then refined by fitting a quadratic interpolation polynomial to the three points making up the bracket and determining the minimum of this polynomial. As a result of the evaluation a new bracket is formed and the procedure repeated. In this case a suitable algorithm has been given by Davies, Swann, and Campey [7, 8].

To bracket the minimum they evaluate $f(x)$ at $x^{(0)}$ and $x^{(0)} + h$. If $f(x^{(0)} + h) \leqslant f(x^{(0)})$, then $f(x)$ is evaluated at $x^{(0)} + 2h$, $x^{(0)} + 4h$, $\ldots$, $x^{(0)} + 2^i h$ until a value of i is found for which $f(x^{(0)} + 2^{(i-1)}h) < f(x^{(0)} + 2^i h)$; otherwise $f(x)$ is evaluated at $x^{(0)} - h$, $x^{(0)} - 2h$, $\ldots$, $x^{(0)} - 2^i h$ until a value of i is found for which $f(x^{(0)} - 2^{(i-1)}h) < f(x^{(0)} - 2^i h)$. The final three points in the computation constitute the bracket. Before predicting the minimum they halve the last step, and from the original bracket and this new point they form a new bracket in which the points are equidistant with separation L (say). If these points are x_1, x_2, and x_3 with $x_1 < x_2 < x_3$, and the corresponding values of $f(x)$ are f_1, f_2, and f_3, then the minimum is predicted to be at

$$x_4 = x_2 - \frac{L}{2} \frac{f_3 - f_1}{f_3 - 2f_2 + f_1}. \tag{2.3.1}$$

This new point is used to produce a new bracket and an additional function value is calculated so that the points in this new bracket are equidistant. The process is then repeated. A procedure which at every stage ensures that the minimum is bracketed, and that the width of the bracket is decreasing can be expected to be very safe in operation. It is likely to require more function values than the approach (i) in circumstances when both work well.

To estimate the rate of convergence of the procedure of successive interpolation we write x_1, x_2, and x_3 for the points on which the interpolation is based. Then we can write $f(x)$ using the Lagrangian interpolation formula with remainder

$$f(x) = \sum_{i=1}^{3} f_i \frac{\prod_{j \neq i}(x - x_j)}{\prod_{j \neq i}(x_i - x_j)} + R_3(x), \tag{2.3.2}$$

where

$$R_3(x) = \tfrac{1}{6} f^{(3)}(\xi(x))(x - x_1)(x - x_2)(x - x_3), \tag{2.3.3}$$

and $\xi(x)$ is a mean value in the interpolation interval. Let x_c satisfy $f^{(1)}(x_c) = 0$.

We compute an approximation to x_c, which we denote by x_4, by differentiating the interpolation polynomial, setting this equal to zero, and solving the resulting linear equation. This gives

$$x_4 = \frac{\dfrac{f_1(x_2 + x_3)}{(x_1 - x_2)(x_1 - x_3)} + \dfrac{f_2(x_1 + x_3)}{(x_2 - x_1)(x_2 - x_3)} + \dfrac{f_3(x_1 + x_2)}{(x_3 - x_1)(x_3 - x_2)}}{2\left\{\dfrac{f_1}{(x_1 - x_2)(x_1 - x_3)} + \dfrac{f_2}{(x_2 - x_1)(x_2 - x_3)} + \dfrac{f_3}{(x_3 - x_1)(x_3 - x_2)}\right\}}, \tag{2.3.4}$$

whence

$$x_c - x_4 = \cfrac{\dfrac{1}{2}\dfrac{dR_3}{dx}(x_c)}{\left\{\dfrac{f_1}{(x_1 - x_2)(x_1 - x_3)} + \dfrac{f_2}{(x_2 - x_1)(x_2 - x_3)} + \dfrac{f_3}{(x_3 - x_1)(x_3 - x_2)}\right\}}$$

(2.3.5)

To compute dR_3/dx we use a result given in Ralston [9]. This is that

$$\frac{1}{n!}\frac{d}{dx}f^{(n)}(\xi(x)) = \frac{1}{(n+1)!}f^{(n+1)}(\eta(x)),$$

(2.3.6)

where η is again a mean value in the interval of interpolation. We also write $x_c - x_i = e_i$, $i = 1, 2, 3, 4$. Then we have

$$e_4(-f_1(e_2 - e_3) - f_2(e_3 - e_1) - f_3(e_1 - e_2))$$

$$= \frac{1}{2}\frac{dR_3}{dx}(x_c)(e_1 - e_2)(e_2 - e_3)(e_3 - e_1) \quad (2.3.7)$$

We write

$$f_i = f(x_c) + \frac{e_i^2}{2}f^{(2)}(x_c) + \mathcal{O}(e_i^3),$$

and assume that third-order terms in e_i can be ignored in this expansion. This gives, on collecting terms,

$$e_4 = \frac{1}{f^{(2)}(x_c)}\frac{dR_3}{dx}(x_c).$$

(2.3.8)

Now

$$\frac{dR_3}{dx} = \tfrac{1}{6}f^{(3)}(\xi)(e_1e_2 + e_3e_2 + e_3e_1) + \tfrac{1}{24}f^{(4)}(\eta)e_1e_2e_3.$$

(2.3.9)

If we assume that $|e_1| > |e_2| > |e_3|$, and that all are small, which will be true if x_1, x_2, and x_3 are increasingly good approximations to the minimum, then e_4 will be approximately given by

$$e_4 = \frac{f^{(3)}(\xi(x_c))}{6f^{(2)}(x_c)}e_1e_2.$$

(2.3.10)

Close enough to the minimum it may be expected that information about the rate of convergence can be obtained by studying the difference equation

$$e_{i+1} = Ke_{i-2}e_{i-1}.$$

(2.3.11)

Making the transformations $E_i = Ke_i$, $d_i = \log E_i$, this becomes

$$d_{i+1} = d_{i-1} + d_{i-2}, \qquad (2.3.12)$$

and this equation has solutions of the form

$$d_i = Ar_1^i + Br_2^i + Cr_3^i, \qquad (2.3.13)$$

where r_1, r_2, r_3 are roots of the equation

$$r^3 - r - 1 = 0. \qquad (2.3.14)$$

This equation has one real root at about $r_1 = 1.3$ and two complex roots (r_2 and r_3), both of which are less than one in modulus. Thus we will have eventually that

$$d_i = -p^2 r_1^i, \qquad (2.3.15)$$

where the coefficient of r_1 takes into account that $d_i \to -\infty$ as $e_i \to 0$. We have

$$e_i = K^{-1} \exp(-p^2 r_1^i), \qquad (2.3.16)$$

whence

$$\frac{e_i}{e_{i-1}^s} = K^{(s-1)} \exp(-p^2 r_1^{i-1}(r_1 - s)), \qquad (2.3.17)$$

and the right-hand side of this equation is independent of i if $s = r_1$. This gives

$$e_i = K^{(r_1-1)} e_{i-1}^{r_1}, \qquad (2.3.18)$$

which shows that convergence is approximately of order 1.3. In particular this means that ultimately the repeated interpolation procedure will converge more rapidly than the algorithm based on the use of Fibonacci numbers, provided the assumption that $f(x)$ can be approximated by a parabola in a neighborhood of x_c is justified. We have used this assumption explicitly because $f^{(2)}(x_c)$ appears in the denominator of the expression for estimating e_4.

To conclude this section we note that if additional information is available then more accurate interpolants might be used. An example occurs in the general minimization techniques that makes use of derivatives. With these, the Hermite interpolation formulas are available, and a suitable procedure is described by Fletcher and Powell [10] and Fletcher and Reeves [11]. The latter also give an ALGOL algorithm. With techniques for solving the general minimization problem it is not obvious that we have to compute the minimum in each search along a line with great accuracy. It is possible that a point giving a lower function value is all that is required. For this procedure we offer the justification that, as the solution is approached, the steps we take shrink, so that eventually a single step of quadratic interpolation should prove adequate to determine well the minimum in each search along a line.

2.4. THE METHOD OF HOOKE AND JEEVES

The direct search method of Hooke and Jeeves [12, 13] is a sequential technique each step of which comprises two kinds of moves: *Exploratory* and *Pattern*. The first kind of move is designed to explore the local behavior of the objective function. Introducing a starting point $\mathbf{x}$, we prescribe step lengths Δx_i in each of the directions $\mathbf{e}_i$, $i = 1, 2, \ldots, N$. The exploratory stage is performed as follows:

(i) Set $i = 1$ and compute $F = f(\mathbf{x})$ where $\mathbf{x} = (x_1, x_2, \ldots, x_N)$.
(ii) Compute the new trial point

$$\mathbf{x} := (x_1, x_2, \ldots, x_i + \Delta x_i, x_{i+1}, \ldots, x_N).$$

(iii) If $f(\mathbf{x}) < F$ then set $F = f(\mathbf{x})$, $i := i + 1$; accept this trial point as a starting point and repeat from (ii).
(iv) If $f(\mathbf{x}) \geqslant F$ then set $\mathbf{x} := (x_1, x_2, \ldots, x_i - 2\Delta x_i, \ldots, x_N)$ and check if $f(\mathbf{x}) < F$. In the case of success the new trial point is retained. We set $F = f(\mathbf{x})$ and repeat from (ii) with $i := i + 1$.
If again $f(\mathbf{x}) \geqslant F$ then the move is rejected, so that x_i remains unchanged and we repeat from (ii) with the new variable, i.e., $i := i + 1$.

The point $\mathbf{x}^B$ arrived at by these Exploratory moves is called a base point. (By definition the starting point is a starting base.) The second type of move is called the Pattern move and it is a simple step from the current base to the point

$$\mathbf{x} = \mathbf{x}^B + (\mathbf{x}^B - \bar{\mathbf{x}}^B) = 2\mathbf{x}^B - \bar{\mathbf{x}}^B, \tag{2.4.1}$$

where $\bar{\mathbf{x}}^B$ is the previous base and $\mathbf{x}^B$ is the current one. Now we *do not* test the function value at $\mathbf{x}$, but start the Exploratory stage again. If the point obtained by the Exploratory moves is better (i.e., the function value is less than at the last base) then it becomes the new base and the process is recommenced. Otherwise, we return to the last base, which becomes a starting base and the process is restarted from it. It is necessary to add that every starting base is treated in a different manner to the usual base point. If the Exploratory moves from any starting base do not yield a point which is better than this base, then we reduce the lengths of all the steps and try again. Convergence is assumed when the step lengths Δx_i have been reduced below predetermined limits.

A simple hypothetical, two-dimensional direct search is presented in Figure 2.5.1. The points have numbers according to the sequence in which they

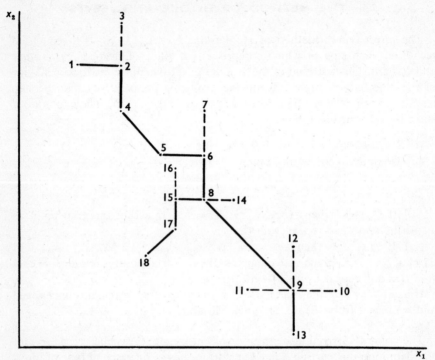

Figure 2.5.1.   Hypothetical two-dimensional direct search of Hooke and Jeeves.

are selected and the function values computed. The point x^1 is a starting base. The new base $x^4 (f(x^4) < f(x^1))$ is obtained by the Exploratory moves where x^3 is a failure and x^2, x^4 are successes. Point x^5 is obtained by the Pattern move. From x^5 we perform again the Exploratory moves, and x^8 becomes a base if $f(x^8) < f(x^4)$. The point x^9 is reached by the Pattern move from x^8. Point x^{13} is the result of the Exploratory moves, but it is not accepted as a base if we assume that $f(x^{13}) > f(x^8)$. We have to return to the last base x^8, which becomes a starting base and recommence the procedure with reduced step lengths.

2.5. THE METHOD OF ROSENBROCK

The method of Rosenbrock [14] can be regarded as a further development of the Hooke and Jeeves technique. It uses a set of n mutually orthogonal directions in each cycle of searches which are similar to Exploratory moves.
Let us introduce the following notation and definitions:

(1) D is an $n \times n$ matrix defining an initial set of n mutually orthogonal directions, for example, the coordinate directions. Let

$$\kappa_i(D) = \mathbf{d}_i \qquad \text{and} \qquad \|\mathbf{d}_i\| = 1.$$

(2) $\{s_i\}$, $i = 1, 2, \ldots, n$ is a set of initial steps to be taken in the corresponding directions $\mathbf{d}_i$.
(3) $\alpha > 1$ and $0 < \beta < 1$ are two preassigned multipliers which are used to control the lengths of the steps.
We say that a step is successful if the new value of the objective function $f(\mathbf{x})$ is less than or equal to the old one.

We shall describe fully a single stage of the Rosenbrock method. We can assume without loss of generality that the stage considered is the first one.

(i) A step s_1 is taken in the direction $\mathbf{d}_1$, from the given starting point $\mathbf{x}^{(0)}$.
(ii) If the step is successful, s_1 is multiplied by α, the new point is retained, and a success is recorded.
(iii) Otherwise the step is retracted, s_1 is multiplied by $-\beta$, and a failure is recorded.
(iv) The above procedure is repeated in the same manner for each direction $\mathbf{d}_i$.

REMARK.   Unlike the Exploratory moves of the Hooke and Jeeves method we repeat (i)–(iv) until at least one trial has been successful and one trial failed in every direction.

Suppose that λ_i is the sum of the successful steps in the directions $\mathbf{d}_i$, $i = 1, 2, \ldots, n$. We now define the following set of vectors

(v)
$$Q = D \begin{bmatrix} \lambda_1 & & & \\ \lambda_2 & \lambda_2 & & \\ \hline & & & \\ \hline \lambda_n & \lambda_n & ------ & \lambda_n \end{bmatrix} \qquad (2.5.1)$$

In this set, $Q_1 = \kappa_1(Q)$ is the vector joining the starting point $x^{(0)}$ and the final point given by the series of searches in the current stage. Vector Q_2 is the sum of the successful steps in all directions, except the first one, and so on. The vectors Q_i are linearly independent and can be used to generate a new set of orthogonal directions. These are obtained by the Gram-Schmidt orthogonalization procedure which gives

(vi) $$V_i = Q_i - \sum_{j=1}^{i-1} (Q_i^T d_j) d_j, \qquad i = 1, 2, \ldots, n, \qquad (2.5.2)$$

where

$$d_i = V_i / \|V_i\|, \qquad (2.5.3)$$

and a new matrix $D = (d_1, d_2, \ldots, d_n)$ is formed.

The orthogonalization process completes the first stage of the method and thereafter the same procedure is repeated until a suitable convergence criterion is satisfied. It is essential that d_1 be in the direction of the total step achieved in the previous stage. It could be expected that this direction is oriented toward the direction of fastest decrease of the function. The direction d_2 is the the most profitable direction that can be constructed normal to d_1, and so on. This feature of the Rosenbrock algorithm is considered a probable reason for its generally satisfactory performance [15]. It has proved very efficient especially in locating early approximations to the minimum.

2.6 THE SIMPLEX METHOD

The simplex method was first described by Himsworth, Spendley, and Hext [16] and later developed by Nelder and Mead [17]. In this method an essential role is played by the geometric figure called a simplex that will be now defined.

DEFINITION. A set of $n + 1$ points in n-dimensional space forms a *simplex*. When the points are equidistant the simplex is said to be *regular*.

In the case $n = 2$ the corresponding figure is an equilateral triangle, while when $n = 3$ it is a tetrahedron. The principal idea of the method is that we can easily form a new simplex from the current one by reflecting one point in the hyperplane spanned by the remaining points. If we choose for this purpose the vertex of the simplex at which the function is greatest, we can expect that at the reflected vertex the function value will be lower. If this is the case then we can continue the process and move our simplex closer to the minimum.

Nelder and Mead suggested a more flexible approach in which the simplex can be altered both in size and in geometry. It loses its regularity but gains the possibility of adapting itself to the local behavior of the objective function.

Let us introduce the following notation:

(1) $\mathbf{x}_h$ is the vertex which corresponds to $f(\mathbf{x}_h) = \max_i f(\mathbf{x}_i)$, where $i = 1, 2, \ldots, n + 1$.

(2) $\mathbf{x}_s$ is the vertex which corresponds to $f(\mathbf{x}_s) = \max_i f(\mathbf{x}_i)$, $i \neq h$.

(3) $\mathbf{x}_l$ is the vertex corresponding to $f(\mathbf{x}_l) = \min_i f(\mathbf{x}_i)$, where $i = 1, 2, \ldots, n + 1$.

(4) $\mathbf{x}_0$ is the centroid of all $\mathbf{x}_i$, $i \neq h$ and is given by

$$\mathbf{x}_0 = \frac{1}{n} \sum_{\substack{i=1 \\ i \neq h}}^{n+1} \mathbf{x}_i. \tag{2.6.1}$$

We now define the three basic operations used in the method:

(1) *Reflection*, where $\mathbf{x}_h$ is replaced by

$$\mathbf{x}_r = (1 + \alpha)\mathbf{x}_0 - \alpha \mathbf{x}_h, \tag{2.6.2}$$

where the reflection coefficient $\alpha > 0$ is equal to the ratio of the distance $[\mathbf{x}_r \mathbf{x}_0]$ to $[\mathbf{x}_h \mathbf{x}_0]$.

(2) *Expansion*, where $\mathbf{x}_r$ is expanded in the direction along which a further improvement of the function value is expected. We use the relation

$$\mathbf{x}_e = \gamma \mathbf{x}_r + (1 - \gamma)\mathbf{x}_0, \tag{2.6.3}$$

where the expansion coefficient $\gamma > 1$ is the ratio of the distance $[\mathbf{x}_e \mathbf{x}_0]$ to $[\mathbf{x}_r \mathbf{x}_0]$.

(3) *Contraction*, by which we contract the simplex,

$$\mathbf{x}_c = \beta \mathbf{x}_h + (1 - \beta)\mathbf{x}_0, \tag{2.6.4}$$

where the contraction coefficient β is the ratio of the distance $[\mathbf{x}_c \mathbf{x}_0]$ to $[\mathbf{x}_h \mathbf{x}_0]$ and satisfies $0 < \beta < 1$.

As we have mentioned, the method can be viewed as the moving, shrinking, and expanding progress of the simplex method toward the minimum. This motion is accomplished in the following way:

(i) An initial simplex is formed, and the function is evaluated at each of the vertices in order to determine $\mathbf{x}_h$, $\mathbf{x}_s$, $\mathbf{x}_l$, and $\mathbf{x}_0$.

(ii) We first try reflection and evaluate the function at the reflected point $\mathbf{x}_r$.

(iii) If $f(\mathbf{x}_s) \geqslant f(\mathbf{x}_r) \geqslant f(\mathbf{x}_l)$, then we replace $\mathbf{x}_h$ by $\mathbf{x}_r$ and restart the process with the newly formed simplex.

(iv) However, if $f(\mathbf{x}_r) < f(\mathbf{x}_l)$, we may expect that the direction $\mathbf{x}_r - \mathbf{x}_0$ could give us an even lower value of the function if we move further. Therefore we expand our new simplex in this direction. The expansion succeeds if $f(\mathbf{x}_l) > f(\mathbf{x}_e)$, and in this case $\mathbf{x}_h$ is replaced by $\mathbf{x}_e$. In the case of failure $\mathbf{x}_h$ is replaced by $\mathbf{x}_r$, and in either case we restart the process from our new simplex.

(v) If the reflection move (ii) yields $\mathbf{x}_r$ such that $f(\mathbf{x}_h) > f(\mathbf{x}_r) > f(\mathbf{x}_s)$, we replace $\mathbf{x}_h$ by $\mathbf{x}_r$ and make the contracting move. (This replacement is not executed when $f(\mathbf{x}_r) > f(\mathbf{x}_h)$.) After the contracting move we compare $f(\mathbf{x}_h)$ and $f(\mathbf{x}_c)$.

If $f(\mathbf{x}_h) > f(\mathbf{x}_c)$, we consider that the contraction is successful, $\mathbf{x}_h$ is replaced by $\mathbf{x}_c$, and we start from the new simplex.

In a case of failure, i.e., $f(\mathbf{x}_h) \leqslant f(\mathbf{x}_c)$, the last simplex is shrunk about the point of the lowest function value $\mathbf{x}_l$ by the relation

$$\mathbf{x}_i := \tfrac{1}{2}(\mathbf{x}_i + \mathbf{x}_l), \qquad (2.6.5)$$

and we begin from (i).

The stopping criterion suggested by Nelder and Mead is

$$\left\{ \frac{1}{n} \sum_{i=1}^{n+1} (f(\mathbf{x}_i) - f(\mathbf{x}_0))^2 \right\}^{1/2} < \varepsilon, \qquad (2.6.6)$$

where ε is some small preset number.

Box [18] has given an alternative version of the simplex method which searches for the minimum of an n-variable function $f(\mathbf{x})$ and takes into account inequality constraints of the form

$$g_i \leqslant x_i \leqslant h_i, \qquad i = 1, 2, \ldots, n, \qquad (2.6.7)$$

and

$$G_j \leqslant \phi_j(\mathbf{x}) \leqslant H_j, \qquad j = 1, 2, \ldots, m, \qquad (2.6.8)$$

where g_i, h_i, G_j, H_j are either constants or functions of $\mathbf{x}$. This variant is called the complex method. The method has the following features:

(1) It uses a nonregular complex with $k > n + 1$ points instead of $k = n + 1$ as in the simplex method. The complex with $k = n + 1$ points (vertices) shows a tendency to collapse into a subspace, particularly when it approaches constraints.

(2) It is assumed that the feasible region is convex and that at least one feasible point is available.

(3) A basic operation used in the complex method is an over-reflection defined by

$$\mathbf{x}_{or} = (1 + \alpha)\mathbf{x}_0 - \alpha\mathbf{x}_h, \qquad (2.6.9)$$

where $\alpha > 1$, $\mathbf{x}_0$ is the centroid, and $\mathbf{x}_h$ is the point with the maximum function value, respectively.

The method works in the following fashion:

(i) Given is an initial feasible point, the further $(k - 1)$ points necessary to set up the initial complex are generated by the formula

$$x_i = g_i + r_i(h_i - g_i), \qquad (2.6.10)$$

where the r_i are pseudo-random numbers evenly distributed in the interval [0, 1]. If a point produced by (2.6.10) does not satisfy constraints (2.6.8), then it is moved half-way towards the centroid of those points which have already been accepted. Ultimately a feasible point will be found.

(ii) The over-reflection move is applied. If the trial point again gives the highest function value then it is moved half-way toward the centroid of the remaining points to give a new trial point. This procedure is repeated until a constraint is violated.

(iii) If over-reflection produces a point which violates an explicit constraint (2.6.7) then this variable is reset to an appropriate boundary value. If some implicit constraint (2.6.8) is violated then the point is moved half-way toward the centroid of the remaining points, and ultimately a feasible point is located.

The stopping criterion suggested by Box is quite simple. The search is stopped when for some consecutive function evaluations the function values differ by less than a prescribed tolerance.

REFERENCES

1. J. Kiefer, Sequential Minimax Search for a Minimum, *Proc. Amer. Math. Soc.*, 4(1953), 502.
2. S. M. Johnson, *Best Exploration for Maximum is Fibonaccian*, The RAND Corporation, RM-1590, 1955.
3. K. J. Overholt, Algorithms Supplement, *Computer J.*, 9(1967), 414.
4. K. J. Overholt, An Instability in the Fibonacci and Golden Section Search Methods, Tidskrift for Informasjons Behandling, 5(1965), 284.
5. J. Boothroyd, Fibonacci Search Algorithms in Theory and Practice, unpublished paper, 1967.
6. M. J. Powell, An Efficient Method of Finding the Minimum of a Function of Several Variables without Calculating Derivatives, *Computer J.*, 7(1964), 155.

7. W. H. Swann, *Report on the Development of a New Direct Search Method of Optmization*, I.C.I. Ltd., Central Instrument Laboratory Research Note 64/3, 1964.
8. Notes from the Lectures on Computational Methods, I.C.I. Ltd., Computer Center, Wilton Works, England, 1965.
9. A. Ralston, *A First Course in Numerical Analysis*, McGraw-Hill Book Co., New York, 1965, p. 77.
10. R. Fletcher and M. J. D. Powell, A Rapidly Convergent Descent Method for Minimization, *Computer J.*, 6(1963), 163.
11. R. Fletcher and C. M. Reeves, Function Minimization by Conjugate Gradients, *Computer J.*, 7(1964), 149.
12. R. Hooke and T. A. Jeeves, Direct Search Solution of Numerical and Statistical Problems, *J. Assoc. Comput. Mach.*, 8(1961), 212.
13. M. Bell and M. C. Pike, Remark on Algorithm 178 [E4] Direct Search, *Comm. ACM*, 9(1966), 684.
14. H. H. Rosenbrock, An Automatic Method for Finding the Greatest or Least Value of a Function, *Computer J.*, 3(1960), 175.
15. *Numerical Analysis, An Introduction*, J. Walsh, ed., Academic Press. London and New York, 1966, p. 148.
16. W. Spendley, G. R. Hext, and F. R. Himsworth, Sequential Application of Simplex Designs in Optimization and Evolutionary Operation, *Technometrics*, 4(1962), 441.
17. J. A. Nelder and R. Mead, A Simplex Method for Function Minimization, *Computer J.*, 7(1965), 308.
18. M. J. Box, A New Method of Constrained Optimization and a Comparison with Other Methods, *Computer J.*, 8(1965), 42.

Chapter 3

DESCENT METHODS

3.1. INTRODUCTION

The methods developed in Chapter 2 for solving general optimization problems use function values only, and use them in a very simple fashion. Generally each value is used to give a yes or no answer to a simple question. While these methods appear to be successful in indicating a region in which a minimum lies, their final rate of convergence is often very disappointing. It is reasonable to expect that more efficient techniques can be devised by utilizing additional information about the function to be optimized. In this chapter we consider a class of such methods in which the solution of the general optimization problem is found by solving a sequence of one-dimensional problems. These methods are called descent methods, and we will often suggest (in Chapter 4 in particular) that techniques be modified to convert them into descent methods as this tactic often improves their global convergence properties.

From the point of view of practical computation the most important descent methods for solving general minimization problems are the *conjugate direction methods*. These assume that in a neighborhood of the minimum the function can be closely approximated by a positive definite quadratic form, and this is the major assumption made in the development of both this and the next chapter.

3.2. DESCENT METHODS

We assume the existence of a local minimum of the function $F(\mathbf{x})$, and we can take this to be at $\mathbf{x} = 0$ without loss of generality. This assumption begs the question of existence, which conceivably could be extremely delicate, but it is no real restriction in practice for it is often the case that the function to be minimized is nonnegative. However, we make further assumptions which are more significant:

(i) The minimum of F at $\mathbf{x} = 0$ is isolated. That is, for $0 < \|\mathbf{x}\| < K_1$ and for K_1 sufficiently small, $F(\mathbf{x}) > F(0)$.

(ii) The function F is strictly convex for $0 \leqslant \|\mathbf{x}\| < K_2$. Again this is assumed to be true only for K_2 small enough.

(iii) The function F possesses continuous partial derivatives to at least third order for $0 \leqslant \|\mathbf{x}\| < K_3$. This implies the inequality (as $\nabla F(0) = 0$)

$$|F(\mathbf{x}) - F(0) - \tfrac{1}{2}\mathbf{x}^T J \mathbf{x}| < K \|\mathbf{x}\|^3, \tag{3.2.1}$$

for $\|\mathbf{x}\| < K_3$ where $J_{pq} = \partial F^2(0)/\partial x_p \, \partial x_q$. By assumption (ii) the matrix J is *positive-definite* (see Appendix 2).

In fact, Appendix 2 shows that the first two assumptions are consequences of assumption (iii) and the positive definiteness of J. However it is instructive to show that (iii) implies (i) directly when J is positive definite. Let $\rho = $ min $\mathbf{t}^T J \mathbf{t}$, $\|\mathbf{t}\| = 1$. That $\rho > 0$ follows because J is positive-definite. Then equation (3.2.1) gives

$$F(\mathbf{x}) > F(0) + \tfrac{1}{2}\mathbf{x}^T J \mathbf{x} - K \|\mathbf{x}\|^3$$
$$> F(0) + \tfrac{1}{2}(\rho - 2K \|\mathbf{x}\|) \|\mathbf{x}\|^2$$
$$> F(0),$$

provided

$$\rho > 2K \|\mathbf{x}\|,$$

and this shows that assumption (i) holds with $K_1 = \rho/2K$.

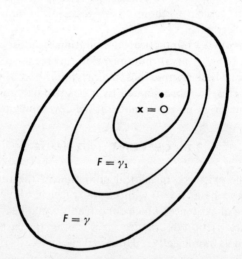

Figure 3.2.1. **In the neighborhood of the origin the level surfaces are closed.**

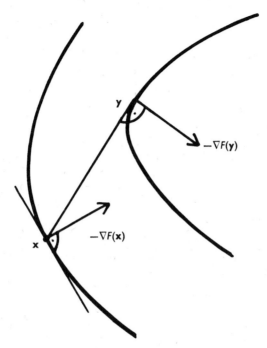

Figure 3.2.2.   The descent vector lies in the tangent plane to the level surface through y.

These assumptions have as a geometric consequence that the level surfaces $F(\mathbf{x}) = \gamma$, for γ close enough to $F(0)$, are closed surfaces enclosing $\mathbf{x} = 0$. Further, if $\gamma_1 < \gamma$ then the surface $F(\mathbf{x}) = \gamma_1$ is enclosed by the surface $F(\mathbf{x}) = \gamma$. In two dimensions a typical situation is sketched in Figure 3.2.1.

In the application of descent methods the following situation is typical. We attempt to move from the current point $\mathbf{x}$ in such a way as to reduce the value of F. Each descent method is characterized by the provision of a descent vector $\mathbf{v}$, and the next point in the interation (say $\mathbf{y}$) is found by solving a one-dimensional minimum problem in which the function to be minimized is $g(\lambda) = F(\mathbf{x} + \lambda\mathbf{v})$.

Geometrically we proceed in the direction of $\mathbf{v}$ (descend!) to lower and lower level surfaces until the point $\mathbf{y}$ is reached. As F cannot be further reduced in this direction $\mathbf{v}$ must lie in the tangent plane to the level surface through $\mathbf{y}$.

$$\therefore \quad \mathbf{v}^T \nabla F(\mathbf{y}) = 0. \tag{3.2.2}$$

The situation is sketched in Figure 3.2.2.

DEFINITION. We say that the vector $\mathbf{v}$ is "downhill" for the function F at the point $\mathbf{x}$ if

$$-\mathbf{v}^T \nabla F(\mathbf{x}) > 0. \tag{3.2.3}$$

We say that a set of unit vectors $\mathbf{v}^{(i)}$ are "*downhill*" for F at the points $\mathbf{x}^{(i)}$ if there exists a δ independent of i such that

$$-\mathbf{v}^{(i)T} \nabla F(\mathbf{x}^{(i)}) / \|\nabla F(\mathbf{x}^{(i)})\| \geqslant \delta > 0. \tag{3.2.4}$$

We can estimate the value of λ which determines the magnitude of the descent step at the current stage of the iteration for we have

$$\begin{aligned}
0 = \frac{dg}{d\lambda} &= \mathbf{v}^T \nabla F(\mathbf{x} + \lambda \mathbf{v}) \\
&= \mathbf{v}^T \{\nabla F(\mathbf{x}) + \lambda \bar{J} \mathbf{v}\},
\end{aligned} \tag{3.2.5}$$

where $J(\mathbf{x})$ is the matrix with components $J_{pq} = \partial^2 F / \partial x_p\, \partial x_q$, and where the bar indicates that mean values of the arguments are appropriate. From equation (3.2.5) the inequality for λ,

$$\delta \, \|\nabla F(x)\| / \|\bar{J}\| \leqslant \lambda, \tag{3.2.6}$$

follows readily. This inequality motivates the choice of step in the iteration which is considered in the following fundamental theorem.

THEOREM 3.2.1 [1]. *Let the closed level surface $F = K$ bound a region R, and let $0 \leqslant F < K$ and $\|J(\mathbf{x})\| \leqslant K_1$ for $\mathbf{x} \in R$. Also let $\nabla F(\mathbf{x}^{(i)}) = k_i\, \mathbf{d}^{(i)}$ where $\|\mathbf{d}^{(i)}\| = 1$, and let the sequences of vectors $\mathbf{x}^{(i)}$ and $\mathbf{v}^{(i)}$ be defined recursively by*

  (i) $\mathbf{x}^{(i+1)} = \mathbf{x}^{(i)} + \dfrac{k_i \delta}{K_1} \mathbf{v}^{(i)}$, *and* $\mathbf{x}^{(1)} \in R$,

  (ii) $\mathbf{v}^{(i)}$ *is "downhill" for F at $\mathbf{x}^{(i)}$,*

  (iii) $\|\mathbf{v}^{(i)}\| = 1$; *then the sequence of numbers $F(\mathbf{x}^{(i)})$ is convergent, and* $\|\nabla F(\mathbf{x}^{(i)})\| \to 0, i \to \infty$.

PROOF. Note first that the iteration is confined to the interior of R. This follows because $\mathbf{v}^{(i)}$ is downhill at each stage, and because (from equation (3.2.6)) each iteration underestimates the step to the minimum in that direction. Thus F is always $< K$ and hence $\mathbf{x}^{(i)}$ cannot escape from R.

By the mean value theorem

$$F(\mathbf{x}^{(i+1)}) = F(\mathbf{x}^{(i)}) + k_i \frac{k_i \delta}{K_1} \mathbf{v}^{(i)T} \mathbf{d}^{(i)} + \frac{1}{2} \left(\frac{k_i \delta}{K_1}\right)^2 \mathbf{v}^{(i)T} \bar{J} \mathbf{v}^{(i)},$$

where the bar denotes that mean values of the arguments of the components are appropriate. As the $\mathbf{v}^{(i)}$ are downhill we have

$$F(\mathbf{x}^{(i+1)}) \leqslant F(\mathbf{x}^{(i)}) - \left(\frac{k_i \delta}{K_1}\right)^2 + \frac{1}{2} \left(\frac{k_i \delta}{K_1}\right)^2 K_1$$

$$\leqslant F(\mathbf{x}^{(i)}) - \frac{1}{2} \frac{(k_i \delta)^2}{K_1}. \tag{3.2.7}$$

Thus the sequence $F(\mathbf{x}^{(i)})$ is decreasing and bounded below. Therefore it is convergent. Whence $|F(\mathbf{x}^{(i+1)}) - F(\mathbf{x}^{(i)})| \to 0$, $i \to \infty$. But, by equation (3.2.7),

$$\|\nabla F(\mathbf{x}^{(i)})\| = k_i \leqslant \frac{1}{\delta} \sqrt{2K_1(F(\mathbf{x}^{(i)}) - F(\mathbf{x}^{(i+1)}))} \to 0, \qquad i \to \infty \tag{3.2.8}$$

and this concludes the proof of the theorem.

REMARK. (i) It requires further assumptions on F before it can be deduced that the $\mathbf{x}^{(i)}$ also tend to a limit. However if J is positive-definite in R then F is strictly convex on R so that there can be at most one point at which $\nabla F = 0$, and this can be the only limit point of the set $\mathbf{x}^{(i)}$.

(ii) As the minimum of F is found in the direction $\mathbf{v}^{(i)}$ in a descent calculation, the inequality (3.2.7) must hold in this case. Thus Ostrowski's theorem also gives information about the convergence of descent methods when the descent vectors $\mathbf{v}^{(i)}$ are downhill.

3.3. RELAXATION

If the search vectors for the descent iteration are chosen to be the coordinate vectors $\mathbf{e}_i$ taken in sequence (with $\mathbf{e}_1$ following $\mathbf{e}_n$) then the resulting method is called *relaxation*. The coordinate vectors used in this way are not necessarily downhill in the sense of the previous section unless ∇F has no zero components throughout the progress of the iteration. However the iteration can stick at a point for a complete cycle of the $\mathbf{e}_i$ only if ∇F is identically zero, that is, if the point is a stationary point for F. If ∇F is available then a set of vectors which are downhill can be generated by using at each stage $\mathbf{e}_s$ where s is the index of the component of ∇F of maximum modulus at the current stage. However the chief advantage of the method is that it does not require ∇F explicitly.

The method of relaxation has been widely used for the iterative solution of sets of linear equations with positive-definite matrix [2]. In this case convergence is guaranteed by the following result.

THEOREM 3.3.1.   *If $A_{ss} > 0$ then the relaxation iteration is convergent for $F = \frac{1}{2}\mathbf{x}^T A\mathbf{x}$ if and only if A is positive-definite.*

REMARK.   If A is positive-definite then $A_{ss} > 0$.

PROOF.   In this case

$$\nabla F = A\mathbf{x}, \tag{3.3.1}$$

so that the displacement at the i-th step of the iteration is determined from the equation

$$\mathbf{e}_s^T A(\mathbf{x}^{(i)} + \lambda \mathbf{e}_s) = 0, \tag{3.3.2}$$

where $s \equiv i \bmod (n)$. This gives

$$\lambda = -(A\mathbf{x}^{(i)})_s / A_{ss}, \tag{3.3.3}$$

whence

$$\begin{aligned} F(\mathbf{x}^{(i+1)}) &= F(\mathbf{x}^{(i)}) + \lambda \mathbf{e}_s^T A\mathbf{x}^{(i)} + \tfrac{1}{2}\lambda^2 A_{ss} \\ &= F(\mathbf{x}^{(i)}) - \tfrac{1}{2}\lambda^2 A_{ss}. \end{aligned} \tag{3.3.4}$$

Now if A is positive-definite then equation (3.3.4) shows that the sequence $F(\mathbf{x}^{(i)})$ is decreasing and bounded below so that it is convergent.

Further, zero must be the limit point of the sequence as the value of F can remain unaltered throughout a complete cycle of the $\mathbf{e}_s$ only if $A\mathbf{x}$ is zero, and this implies $\mathbf{x}$ is zero as A is positive-definite. However if A is not positive-definite then there is an $\mathbf{x}$ such that $F(\mathbf{x}) < 0$, and equation (3.3.4) shows that the relaxation iteration cannot reduce the magnitude of $F(\mathbf{x})$ in this case.

REMARK.   (i) Presumably information concerning the ultimate convergence of the relaxation iteration can be obtained by expanding F in a Taylor series up to terms of second order. This result then underlines the reasonableness of our initial assumptions (Section 2) in this case.

(ii) The rate of convergence of the relaxation iteration for the solution of systems of linear equations has been studied in great detail in certain cases, and a strategy of systematically over-correcting (over-relaxation) has proved highly successful [3]. There appears to be little information available when F is not a positive-definite quadratic form.

3.4. THE METHOD OF STEEPEST DESCENT

Another possibility that springs to mind is the use of $-\nabla F$ as the descent vector at each stage. Certainly this vector is downhill, and, further, no vector can give a greater local reduction in F as the gradient vector cuts the adjacent level surfaces at right angles. The resulting method is called *the method of steepest descent*.

The geometric observation made in the previous paragraph has caused considerable interest to be shown in this method. However the results have not, in general, justified the energy expended in obtaining them. (This shows that a good local strategy is not necessarily a good global strategy.) The general pattern of observation has been that after an initial satisfactory phase the iteration gets into a "cage" [4] and subsequently makes only slow progress—possibly even giving apparent convergence if an inappropriate test is used.

An explanation for this behavior that is sometimes offered is that the steepest descent computation is unstable with respect to small perturbations. The situation assumed is that the steepest descent direction is a favorable one but that the directions at points close by are not. Thus errors creeping in through rounding error and through failure to compute the precise minimum in the previous descent step mean that the direction for the next step is likely to be unfavorable.

To obtain some insight into this possibility, consider what is probably the worst case—when the direction at the perturbed point is orthogonal to the direction at the point given by precise computation. This gives

$$\nabla F(\mathbf{x}) \cdot \nabla F(\mathbf{x} + \delta\mathbf{x}) = 0, \qquad (3.4.1)$$

whence, by the mean value theorem,

$$\|\nabla F(\mathbf{x})\|^2 + \nabla F(\mathbf{x}) \cdot J\,\delta\mathbf{x} = 0. \qquad (3.4.2)$$

Thus

$$\|\delta\mathbf{x}\| \geqslant \frac{\|\nabla F(\mathbf{x})\|}{\|J\|}, \qquad (3.4.3)$$

which gives a lower bound for the assumed perturbation $\|\delta\mathbf{x}\|$ in terms of quantities that can be assumed to be evaluated at $\mathbf{x}$ provided $\|\delta\mathbf{x}\|$ is small enough. Equation (3.4.3) characterizes the situation as one in which the minimum is poorly determined so that $\|\nabla F\|$ is small in a region containing the minimum, and the level surfaces are sharply curved so that $\|J\|$ is large. A typical situation is one where the minimum is approached along the fairly level floor of a steep valley, and is sketched in Figure 3.4.1.

However this is not adequate to explain why steepest descent is a fairly consistently poor computational strategy for problems in which no particular difficulty is anticipated, as both equation (3.4.3) and Figure 3.4.1 refer to extreme cases. It is more likely that the difficulty lies in the assumption that the steepest descent direction is a good one. In this connection, we note that when the method of steepest descent is used to minimize a function of two

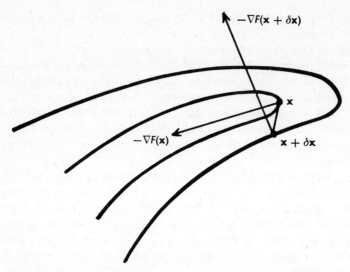

**Figure 3.4.1.   Illustration of a difficulty with the method
of steepest descent.**

variables the directions of descent are fixed by the choice of the initial point
to be opposite to the initial gradient vector and perpendicular to it. As only
these two directions are used, the method of steepest descent is identical (for
two variables) with the method of relaxation applied with suitably rotated
axes (the rotation being a function of the initial point only).

Even more interesting are the results of Akaike [5], who proves a con-
jecture of Forsythe and Motzkin [6] on the ultimate nature of the convergence
of the method of steepest descent applied to a positive-definite quadratic
form. He proves that the directions used are ultimately asymptotic to just
two directions so that eventually the minimum is approached in a two-
dimensional subspace. This would appear to indicate the major defect of the
method, and it is likely that something like this happens also in the general
case (according to Forsythe [7] these directions form the cage of Stiefel.)

In the case in which F is a positive-definite quadratic form then the ultimate
rate of convergence of the method of steepest descent is at least geometric.
It is sufficient to consider $F = \frac{1}{2}\mathbf{x}^T A\mathbf{x}$, and in this case

$$\nabla F = A\mathbf{x},$$

so that

$$\mathbf{x}^{(i+1)} = \mathbf{x}^{(i)} - \gamma A\mathbf{x}^{(i)}$$

$$= \frac{P_1(A)}{P_1(0)} \mathbf{x}^{(i)}, \tag{3.4.4}$$

where $P_1(\lambda) = 1 - \gamma\lambda$, and we have

$$F(\mathbf{x}^{(i+1)}) = F\left(\frac{P_1(A)}{P_1(0)}\mathbf{x}^{(i)}\right)$$

$$\leqslant F\left(\frac{Q_1(A)}{Q_1(0)}\mathbf{x}^{(i)}\right), \tag{3.4.5}$$

where Q_1 is any polynomial of degree 1 in A.

We specialize Q_1 to be the polynomial of degree 1 defined by

$$Q_1(\lambda_1) = -1,$$
$$Q_1(\lambda_n) = 1,$$

where λ_1 and λ_n are the least and greatest eigenvalues of the matrix A. As A is positive-definite all its eigenvalues are positive. We have

$$Q_1(t) = \{2t - (\lambda_1 + \lambda_n)\}/(\lambda_n - \lambda_1). \tag{3.4.6}$$

Let the eigenvalues of A be $\lambda_1, \lambda_2, \ldots, \lambda_n$, and let the corresponding eigenvectors be $\mathbf{v}_i$, $i = 1, 2, \ldots, n$. Then the current point can be written

$$\mathbf{x}^{(i)} = \sum_{j=1}^{n} a_j^{(i)}\mathbf{v}_j,$$

and the bound for $F(\mathbf{x}^{(i+1)})$ becomes

$$F(\mathbf{x}^{(i+1)}) \leqslant \frac{1}{2}\frac{1}{Q_1(0)^2}\sum_j a_j^{(i)2}Q_1(\lambda_j)^2\lambda_j$$

$$\leqslant \frac{1}{2}\frac{1}{Q_1(0)^2}\sum_j a_j^{(i)2}\lambda_j,$$

as $|Q_1(\lambda_j)| \leqslant 1, j = 1, 2, \ldots, n$, by the definition of Q_1, whence

$$F(\mathbf{x}^{(i+1)}) \leqslant \frac{1}{Q_1(0)^2}F(\mathbf{x}^{(i)})$$

$$\leqslant \left(\frac{\lambda_n - \lambda_1}{\lambda_n + \lambda_1}\right)^2 F(\mathbf{x}^{(i)}), \tag{3.4.7}$$

which establishes the stated result. Actually the rate of convergence is exactly geometric, and a proof of the complementary inequality is given in Forsythe [7].

3.5. CONJUGATE DIRECTIONS FOR MINIMIZING A QUADRATIC FORM

In the preceding section it was indicated that the method of steepest descent ultimately approached a minimum in a two-dimensional subspace, and we suggested that this, rather than any apparent instability of the method, was the reason for its generally poor performance. This suggests that a good method will have the characteristic that any t consecutive search directions ($t \leqslant n$) are linearly independent. This is true (at least ultimately) for the class of methods based on the use of conjugate directions, and this family of methods provides us with the most useful general-purpose minimization methods currently available. The description of this class of methods is the subject of the remaining sections of this chapter.

DEFINITION.   Two directions $\mathbf{u}$ and $\mathbf{v}$ are said to be conjugate with respect to the positive-definite matrix A if

$$\mathbf{u}^T A \mathbf{v} = 0. \tag{3.5.1}$$

Where no possibility of confusion arises, explicit reference will not be made to the matrix, and one will simply say that $\mathbf{u}$ and $\mathbf{v}$ are conjugate.

REMARK.   There exists at least one set of n independent vectors mutually conjugate with respect to A as the eigenvectors of A form such a set.

It is possible to generate a set of conjugate directions by a process analogous to the Gram-Schmidt orthogonalization procedure. For example, if we start with a vector $\mathbf{d}_1$, then we can define

$$\mathbf{d}_2 = A\mathbf{d}_1 - b_{11}\mathbf{d}_1, \tag{3.5.2}$$

and $\mathbf{d}_2$ will be conjugate to $\mathbf{d}_1$ if

$$b_{11} = \frac{\mathbf{d}_1^T A^2 \mathbf{d}_1}{\mathbf{d}_1^T A \mathbf{d}_1} \tag{3.5.3}$$

Proceeding in similar fashion we set

$$\mathbf{d}_3 = A\mathbf{d}_2 - b_{22}\mathbf{d}_2 - b_{21}\mathbf{d}_1, \tag{3.5.4}$$

whence

$$b_{22} = \frac{\mathbf{d}_2^T A^2 \mathbf{d}_2}{\mathbf{d}_2^T A \mathbf{d}_2},$$

and

$$b_{21} = \frac{\mathbf{d}_1^T A^2 \mathbf{d}_2}{\mathbf{d}_1^T A \mathbf{d}_1}$$

$$= \frac{(\mathbf{d}_2 + b_{11}\mathbf{d}_1)^T A \mathbf{d}_2}{\mathbf{d}_1^T A \mathbf{d}_1}.$$

using equation (3.5.2)

$$= \frac{\mathbf{d}_2^T A \mathbf{d}_2}{\mathbf{d}_1^T A \mathbf{d}_1}. \tag{3.5.5}$$

It is now readily established by induction that

$$\mathbf{d}_{i+1} = A\mathbf{d}_i - \frac{\mathbf{d}_i^T A^2 \mathbf{d}_i}{\mathbf{d}_i^T A \mathbf{d}_i}\mathbf{d}_i - \frac{\mathbf{d}_i^T A \mathbf{d}_i}{\mathbf{d}_{i-1}^T A \mathbf{d}_{i-1}}\mathbf{d}_{i-1}, \tag{3.5.6}$$

for if we assume that this relation is valid for $j = 1, 2, \ldots, i$, and that

$$\mathbf{d}_{i+1} = A\mathbf{d}_i - \sum_{s=1}^{i} b_{is}\mathbf{d}_s$$

then $b_{is} = 0$ if $s < i - 1$ so that the relation holds also for $j = i + 1$. We have

$$0 = \mathbf{d}_s^T A \mathbf{d}_{i+1} = \mathbf{d}_s^T A^2 \mathbf{d}_i - \mathbf{d}_s^T A \mathbf{d}_s b_{is},$$

and

$$\mathbf{d}_s^T A^2 \mathbf{d}_i = (\mathbf{d}_{s+1} + b_{s(s-1)}\mathbf{d}_{s-1})^T A \mathbf{d}_i,$$

by the three-term relation assumed to hold for $s \leqslant i - 1$. This vanishes for $s < i - 1$, as $\mathbf{d}_{s+1}$ and $\mathbf{d}_{s-1}$ are conjugate to $\mathbf{d}_i$ if $s < i - 1$. This shows that $b_{is} = 0$ if $s < i - 1$.

If vectors $\mathbf{d}_1, \mathbf{d}_2, \ldots, \mathbf{d}_n$ are mutually conjugate then clearly they are linearly independent. Therefore any vector conjugate to all of them must be the zero vector. In particular this means that $\mathbf{d}_{n+1}$ vanishes identically in the above scheme.

The significance of conjugate directions for our purposes is contained in the following result.

THEOREM 3.5.1. *If $\mathbf{d}_1, \mathbf{d}_2, \ldots, \mathbf{d}_n$ are a set of vectors mutually conjugate with respect to the positive-definite matrix A then the minimum of the quadratic form*

$$F = a + \mathbf{b}^T \mathbf{x} + \tfrac{1}{2}\mathbf{x}^T A \mathbf{x}, \tag{3.5.7}$$

can be found from an arbitrary starting point $\mathbf{x}^{(0)}$ *by a finite descent computation in which each of the vectors* $\mathbf{d}_i$ *is used as a descent direction only once. The order in which the* $\mathbf{d}_i$ *are used is immaterial.*

PROOF. We note that as the $\mathbf{d}_i$ are linearly independent an arbitrary vector $\mathbf{v}$ can be written in the form

$$\mathbf{v} = \sum_{i=1}^{n} \alpha_i \mathbf{d}_i, \tag{3.5.8}$$

where

$$\alpha_i = \frac{\mathbf{d}_i^T A \mathbf{v}}{\mathbf{d}_i^T A \mathbf{d}_i}, \tag{3.5.9}$$

by the conjugacy of the vectors $\mathbf{d}_i$.

Let us assume that at the p-th stage of the calculation we have reached the point

$$\mathbf{x}^{(p)} = \mathbf{x}^{(0)} + \sum_{i=1}^{p} \lambda_i \mathbf{d}_i. \tag{3.5.10}$$

We move from $\mathbf{x}^{(p)}$ to $\mathbf{x}^{(p+1)}$ along the direction $\mathbf{d}_{p+1}$, and the value of λ_{p+1} is given by the equation

$$\mathbf{d}_{p+1}^T \nabla F(\mathbf{x}^{(p+1)}) = 0. \tag{3.5.11}$$

This gives

$$\mathbf{d}_{p+1}^T \left(A \left\{ \mathbf{x}^{(0)} + \sum_{i=1}^{p} \lambda_i \mathbf{d}_i + \lambda_{p+1} \mathbf{d}_{p+1} \right\} + \mathbf{b} \right) = 0, \tag{3.5.12}$$

so that

$$\lambda_{p+1} = - \frac{\mathbf{d}_{p+1}^T (A\mathbf{x}^{(0)} + \mathbf{b})}{\mathbf{d}_{p+1}^T A \mathbf{d}_{p+1}}, \tag{3.5.13}$$

which depends only on the starting position $\mathbf{x}^{(0)}$ and not on the path by which $\mathbf{x}^{(p+1)}$ is reached from $\mathbf{x}^{(0)}$.

After n steps we have

$$\mathbf{x}^{(n)} = \mathbf{x}^{(0)} - \sum_{i=1}^{n} \frac{\mathbf{d}_i^T (A\mathbf{x}^{(0)} + \mathbf{b}) \mathbf{d}_i}{\mathbf{d}_i^T A \mathbf{d}_i}, \tag{3.5.14}$$

and, by equations (3.5.8) and (3.5.9), this is equivalent to

$$\begin{aligned} \mathbf{x}^{(n)} &= \mathbf{x}^{(0)} - \mathbf{x}^{(0)} - A^{-1}\mathbf{b} \\ &= -A^{-1}\mathbf{b}, \end{aligned} \tag{3.5.15}$$

which shows that the minimum has been reached.

3.6. CONJUGATE DIRECTIONS FOR THE GENERAL OPTIMIZATION PROBLEM

A general theme in the presentation of this chapter is that considerable information about the ultimate performance of a method can be obtained by considering it applied to a positive-definite quadratic form. The feasibility of this procedure stems from our main assumption that the function to which the optimization technique is applied is adequately represented by a quadratic form in the neighborhood of the sought minimum. Descent methods using conjugate directions are strictly applicable only to quadratic forms so that to use them for minimizing a general function we require a rule for generating directions of search which produces conjugate directions when the function is specialized to a quadratic form. In view of our main assumption a method of this kind can be expected to have a fast rate of ultimate convergence because of the finite nature of the iteration when applied to a quadratic form, and we say these methods are *quadratically convergent*.

A method for generating conjugate directions that will apply also to more general functions must be able to be expressed in terms of quantities such as function values and gradient vectors, and it must not assume explicitly the matrix of the quadratic form. The following lemma is important in developing the methods of this and the next section.

LEMMA 3.6.1. *Let the point* $\mathbf{x}^{(i+1)}$ *be reached by i descent steps applied to the quadratic form* (3.5.7) *where the descent directions* $\mathbf{d}_1, \mathbf{d}_2, \ldots, \mathbf{d}_i$ *are mutually conjugate with respect to the matrix A, then*

$$\mathbf{d}_s^T \mathbf{g}_{i+1} = 0, \qquad s = 1, 2, \ldots, i, \tag{3.6.1}$$

where

$$\mathbf{g}_{i+1} = \nabla F(\mathbf{x}^{(i+1)}).$$

REMARK. This result permits an easy proof of Theorem 3.5.1 for after n steps $\mathbf{g}_{n+1}$ is orthogonal to n linearly independent vectors so that it must vanish identically.

PROOF. We have

$$\mathbf{g}_{i+1} = A\mathbf{x}^{(i+1)} + \mathbf{b}$$

$$= A\left(\mathbf{x}^{(s+1)} + \sum_{j=s+1}^{i} \lambda_j \mathbf{d}_j\right) + \mathbf{b}$$

$$= \mathbf{g}_{s+1} + \sum_{j=s+1}^{i} \lambda_j A\mathbf{d}_j,$$

so that

$$\mathbf{d}_s^T \mathbf{g}_{i+1} = \mathbf{d}_s^T \mathbf{g}_{s+1} + \sum_{j=s+1}^{i} \lambda_j \mathbf{d}_s^T A \mathbf{d}_j$$
$$= 0,$$

because

(i) $\mathbf{d}_s^T \mathbf{g}_{s+1} = 0$ is the condition for descent along $\mathbf{d}_s$,

(ii) $\sum_{j=s+1}^{i} \lambda_j \mathbf{d}_s^T A \mathbf{d}_j = 0$ by conjugacy.

As a first attempt at developing an algorithm we try to modify the method of steepest descent applied to the quadratic form (3.5.7) by imposing the further condition that the successive descent directions be mutually conjugate. This approach can be motivated by noting that, by lemma 3.6.1, the direction of steepest descent at the current point is linearly independent of the previous descent directions.

We begin by setting

$$\mathbf{d}_1 = -\mathbf{g}_1 = -A\mathbf{x}^{(1)} - \mathbf{b}. \tag{3.6.2}$$

We have $\mathbf{x}^{(2)} = \mathbf{x}^{(1)} + \lambda_1 \mathbf{d}_1$, where λ_1 is found from

$$\mathbf{d}_1^T (A(\mathbf{x}^{(1)} + \lambda_1 \mathbf{d}_1) + \mathbf{b}) = 0, \tag{3.6.3}$$

but it could equally well be found by minimizing $F(\mathbf{x}^{(1)} + \lambda \mathbf{d}_1)$ as a function of λ by the methods of Chapter 2 and this approach is applicable when F is not a quadratic form.

We now set

$$\mathbf{d}_2 = -\mathbf{g}_2 + \gamma_2 \mathbf{d}_1, \tag{3.6.4}$$

where γ_2 is chosen to make $\mathbf{d}_1$ and $\mathbf{d}_2$ conjugate. This gives

$$0 = \mathbf{d}_1^T A \mathbf{d}_2 = -\mathbf{d}_1^T A \mathbf{g}_2 + \gamma_2 \mathbf{d}_1^T A \mathbf{d}_1$$
$$= -(\mathbf{x}^{(2)} - \mathbf{x}^{(1)})^T A(\mathbf{g}_2 - \gamma_2 \mathbf{d}_1)$$
$$= -(\mathbf{g}_2 - \mathbf{g}_1)^T (\mathbf{g}_2 - \gamma_2 \mathbf{d}_1),$$

whence

$$\gamma_2 = \frac{\mathbf{g}_2^T \mathbf{g}_2}{\mathbf{g}_1^T \mathbf{g}_1}, \tag{3.6.5}$$

so that it can be evaluated also for a general function. For the next step we set

$$\mathbf{d}_3 = -\mathbf{g}_3 + \gamma_3 \mathbf{d}_2 + \delta_3 \mathbf{d}_1, \tag{3.6.6}$$

and we consider the coefficient δ_3 (a general expression for γ_i will be derived subsequently). We have

$$\delta_3 = \frac{\mathbf{d}_1^T A \mathbf{g}_3}{\mathbf{d}_1^T A \mathbf{d}_1}$$

$$= \frac{1}{\lambda_1} \frac{(\mathbf{x}^{(2)} - \mathbf{x}^{(1)})^T A \mathbf{g}_3}{\mathbf{d}_1^T A \mathbf{d}_1}$$

$$= \frac{1}{\lambda_1} \frac{(\mathbf{g}_2 - \mathbf{g}_1)^T \mathbf{g}_3}{\mathbf{d}_1^T A \mathbf{d}_1}$$

$$= 0,$$

as $\mathbf{g}_2 - \mathbf{g}_1 = -\mathbf{d}_2 + (1 + \gamma_2)\mathbf{d}_1$ is orthogonal to $\mathbf{g}_3$ by Lemma 3.6.1.

This suggests that we attempt to establish inductively a relation of the form

$$\mathbf{d}_i = -\mathbf{g}_i + \gamma_i \mathbf{d}_{i-1}, \tag{3.6.7}$$

which we have shown to hold for $i = 1, 2, 3$. We verify the existence of such a relation in three steps:

(i) On the basis of the induction hypothesis we have

$$\mathbf{g}_s^T \mathbf{g}_i = 0, \qquad s = 1, 2, \ldots, i - 1. \tag{3.6.8}$$

This is a consequence of Lemma 3.6.1 which gives

$$0 = \mathbf{d}_s^T \mathbf{g}_i = (-\mathbf{g}_s + \gamma_s \mathbf{d}_{s-1})^T \mathbf{g}_i = -\mathbf{g}_s^T \mathbf{g}_i.$$

(ii) We have $\mathbf{d}_i$ conjugate to $\mathbf{d}_s$, $s = 1, 2, \ldots, i - 2$, for (using equation (3.6.7))

$$-\mathbf{d}_s^T A \mathbf{d}_i = \mathbf{d}_s^T A \mathbf{g}_i$$

$$= \frac{1}{\lambda_s} (\mathbf{x}^{(s+1)} - \mathbf{x}^{(s)})^T A \mathbf{g}_i$$

$$= \frac{1}{\lambda_s} (\mathbf{g}_{s+1} - \mathbf{g}_s)^T \mathbf{g}_i$$

$$= 0, \qquad s = 1, 2, \ldots, i - 2,$$

by the result (i) above.

(iii) We can choose γ_i so that $\mathbf{d}_i$ is conjugate also to $\mathbf{d}_{i-1}$ (this completes the induction). Further γ_i can be evaluated for a general function. In this case

$$\gamma_i = \frac{\mathbf{d}_{i-1}^T A \mathbf{g}_i}{\mathbf{d}_{i-1}^T A \mathbf{d}_{i-1}}$$

$$= -\frac{(\mathbf{g}_i - \mathbf{g}_{i-1})^T \mathbf{g}_i}{(\mathbf{g}_i - \mathbf{g}_{i-1})^T \mathbf{g}_{i-1}}$$

$$= \frac{\mathbf{g}_i^T \mathbf{g}_i}{\mathbf{g}_{i-1}^T \mathbf{g}_{i-1}}. \tag{3.6.9}$$

We have established the existence of a method for generating conjugate directions which can be applied to a general function provided we can (i) find the minimum of this function along a line, and (ii) evaluate the gradient vector at a point. The use of this method for general functions was suggested by Fletcher and Reeves [8], who found that it is necessary in practice to restart the method every so often by a new steepest descent step. The figure of $(n + 1)$ iterations before restart is suggested. In adopting this figure Fletcher and Reeves are following Hestenes and Stiefel [9], who developed the iteration much earlier for the solution of systems of linear equations. They noted that the method did not terminate due to the cummulative effect of rounding errors so they continued it iteratively. However they noted that the $(n + 1)$-th step was often particularly favorable and suggested the restart procedure on this basis. The reader interested in the use of this and similar methods for the solution of systems of linear equations can find additional information and alternative approaches in Stiefel [10] and Beckman [11].

We summarize the algorithm of Fletcher and Reeves as follows:

  (i) Compute $\mathbf{g}_1 = \nabla F(\mathbf{x}^{(1)})$, and set $\mathbf{d}_1 = -\mathbf{g}_1$.
 (ii) For $i = 2, 3, \ldots, n + 2$
    (a) set $\mathbf{x}^{(i)} = \mathbf{x}^{(i-1)} + \lambda_{i-1}\mathbf{d}_{i-1}$, where
    (b) λ_{i-1} minimizes $q_{i-1}(\lambda) = F(\mathbf{x}^{(i-1)} + \lambda_{i-1}\mathbf{d}_{i-1})$;
    (c) compute $\mathbf{g}_i = \nabla F(\mathbf{x}^{(i)})$;
    (d) unless $i = n + 2$ set

$$\mathbf{d}_i = -\mathbf{g}_i + \frac{\mathbf{g}_i^T \mathbf{g}_i}{\mathbf{g}_{i-1}^T \mathbf{g}_{i-1}} \mathbf{d}_{i-1}$$

and repeat (a)
 (iii) Replace $\mathbf{x}^{(1)}$ by $\mathbf{x}^{(n+2)}$ and restart the iteration.

3.7. THE METHOD OF DAVIDON, FLETCHER, AND POWELL

As conjugate directions provide a method for solving a set of linear equations in a finite number of steps, it is natural to ask if an explicit form for the inverse matrix can be given in terms of them. Consider the matrix

$$\sum_{i=1}^{p} \alpha_i \mathbf{d}_i \mathbf{d}_i^T$$

where $\mathbf{d}_1, \mathbf{d}_2, \ldots, \mathbf{d}_p$ are p conjugate directions. We have for $s = 1, 2, \ldots, p$

$$\left(\sum_{i=1}^{p} \alpha_i \mathbf{d}_i \mathbf{d}_i^T \right) A\mathbf{d}_s = \alpha_s \mathbf{d}_s \mathbf{d}_s^T A\mathbf{d}_s = \mathbf{d}_s, \tag{3.7.1}$$

provided $\alpha_s = 1/\mathbf{d}_s^T A\mathbf{d}_s$. This gives, in particular, for $p = n$

$$A^{-1} = \sum_{i=1}^{n} \frac{\mathbf{d}_i \mathbf{d}_i^T}{\mathbf{d}_i^T A\mathbf{d}_i}, \tag{3.7.2}$$

and the partial sums of this expression serve as approximations to the inverse in the sense expressed by equation (3.7.1)—that is, they act as projections of the inverse matrix into the subspaces spanned by $\mathbf{d}_1, \ldots, \mathbf{d}_p$ for $p = 1, 2, \ldots, n$.

Now it is obvious that if the inverse matrix is known then the quadratic form (3.5.7) can be minimized in one step. This is illustrated in Figure 3.7.1.

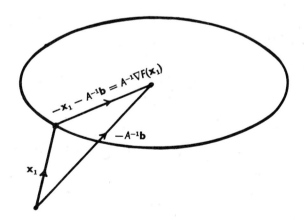

Figure 3.7.1. Minimization of a quadratic form when A^{-1} is known.

These preliminaries suggest an iterative scheme in which the best approximation to the inverse (say H_i) is used to define the next direction of search by

$$\mathbf{d}_{i+1} = -H_i \, \nabla F(\mathbf{x}^{(i+1)}), \tag{3.7.3}$$

and the results of this search are then used to improve the approximation to the inverse. We also ask if it is possible to arrange the successive approximations to the inverse so that the successive directions generated are conjugate.

Thus we seek an algorithm of the form

$$\mathbf{d}_i = -H_{i-1}\mathbf{g}_i,$$

$$\mathbf{x}^{(i+1)} = \mathbf{x}^{(i)} + \lambda_i \mathbf{d}_i,$$

$$H_i = H_{i-1} + \frac{\mathbf{d}_i \mathbf{d}_i^T}{\mathbf{d}_i^T A \mathbf{d}_i} + B_i,$$

where, in updating H_{i-1}, we have used equation (3.7.2) as a model and added a matrix B_i as a correcting term. To derive an appropriate form for B_i we note (bearing in mind equation (3.7.1) and the desirability of the search vectors $\mathbf{d}_i$ being conjugate) that if $\mathbf{d}_1, \ldots, \mathbf{d}_{i-1}$ are conjugate, and if

$$\mathbf{d}_s^T A H_{i-1} = \mathbf{d}_s^T, \qquad s = 1, 2, \ldots, i-1,$$

the $\mathbf{d}_1, \ldots, \mathbf{d}_i$ are conjugate for

$$-\mathbf{d}_s^T A \mathbf{d}_i = \mathbf{d}_s^T A H_{i-1}\mathbf{g}_i = \mathbf{d}_s^T \mathbf{g}_i = 0, \tag{3.7.4}$$

by Lemma 3.6.1. Thus we seek to choose B_i to satisfy

$$\mathbf{d}_s^T A H_i = \mathbf{d}_s^T, \qquad s = 1, 2, \ldots, i, \tag{3.7.5}$$

and we assume the conjugacy of $\mathbf{d}_1, \mathbf{d}_2, \ldots, \mathbf{d}_i$.

For $s = i$, equation (3.7.5) gives

$$\mathbf{d}_i^T A \left(H_{i-1} + \frac{\mathbf{d}_i \mathbf{d}_i^T}{\mathbf{d}_i^T A \mathbf{d}_i} + B_i \right) = \mathbf{d}_i^T,$$

whence

$$(\mathbf{g}_{i+1} - \mathbf{g}_i)^T (H_{i-1} + B_i) = 0,$$

and the simplest solution to this equation is (writing $\mathbf{y}_i = \mathbf{g}_{i+1} - \mathbf{g}_i$)

$$B_i = -\frac{H_{i-1}\mathbf{y}_i\mathbf{y}_i^T H_{i-1}}{\mathbf{y}_i^T H_{i-1}\mathbf{y}_i}. \tag{3.7.6}$$

From this equation it follows immediately that

$$\mathbf{d}_s^T A B_i = 0, \qquad s = 1, 2, \ldots, i-1,$$

by Lemma 3.6.1.

This completes the specification of the algorithm, but to make it applicable to a general function it is necessary to modify the expression for updating H_{i-1} to remove the term $\mathbf{d}_i^T A \mathbf{d}_i$. We have

$$
\begin{aligned}
\mathbf{d}_i^T A \mathbf{d}_i &= \frac{(\mathbf{x}^{(i+1)} - \mathbf{x}^{(i)})^T}{\lambda_i} A \mathbf{d}_i \\
&= \frac{(\mathbf{g}_{i+1} - \mathbf{g}_i)^T}{\lambda_i} \mathbf{d}_i \\
&= \frac{1}{\lambda_i} \mathbf{g}_i^T H_{i-1} \mathbf{g}_i,
\end{aligned}
\tag{3.7.7}
$$

and this shows that the expression can be evaluated also for a general function. We summarize the i-th stage of the algorithm as follows:

(i) Compute $\mathbf{d}_i = -H_{i-1}\mathbf{g}_i$.
(ii) Compute λ_i to minimize

$$q_i(\lambda) = F(\mathbf{x}^{(i)} + \lambda \mathbf{d}_i).$$

(iii) Set $\mathbf{x}^{(i+1)} = \mathbf{x}^{(i)} + \lambda_i \mathbf{d}_i$,

$$\mathbf{y}_i = \mathbf{g}_{i+1} - \mathbf{g}_i$$

(iv) Compute

$$H_i = H_{i-1} + \lambda_i \frac{\mathbf{d}_i \mathbf{d}_i^T}{\mathbf{g}_i^T H_{i-1} \mathbf{g}_i} - \frac{H_{i-1} \mathbf{y}_i \mathbf{y}_i^T H_{i-1}}{\mathbf{y}_i^T H_{i-1} \mathbf{y}_i}.$$

In the first step of the iteration it is customary to set $H_0 = I$ so that the first step is equivalent to a step of the steepest descent method. However, if an estimate of A^{-1} is known this can be used provided it is positive-definite.

If H_0 is positive-definite initially, then it is readily seen inductively that all subsequent H_i are also positive-definite. As $H_n = A^{-1}$ by construction it must be positive-definite so that if H_0 does not satisfy this condition there is the possibility of a breakdown in the calculation (instability). To carry out the induction proof we assume H_{i-1} is positive-definite and prove that this is true also of H_i. On this hypothesis we have

$$\mathbf{d}_i^T \mathbf{g}_i = -\mathbf{g}_i^T H_{i-1} \mathbf{g}_i < 0, \tag{3.7.8}$$

so that $\mathbf{d}_i$ is downhill and, consequently, $\lambda_i > 0$. Now let $\mathbf{u}$ be an arbitrary vector with $\|\mathbf{u}\| \neq 0$, then

$$\mathbf{u}^T H_i \mathbf{u} = \mathbf{u}^T H_{i-1} \mathbf{u} + \lambda_i \frac{(\mathbf{d}_i^T \mathbf{u})^2}{\mathbf{g}_i^T H_{i-1} \mathbf{g}_i} - \frac{(\mathbf{u}^T H_{i-1} \mathbf{y}_i)^2}{\mathbf{y}_i^T H_{i-1} \mathbf{y}_i}. \tag{3.7.9}$$

As $(\mathbf{u}^T H_{i-1} \mathbf{y}_i)^2 \leqslant (\mathbf{u}^T H_{i-1} \mathbf{u})(\mathbf{y}_i^T H_{i-1} \mathbf{y}_i)$ with equality only if $\mathbf{u}$ is parallel to $\mathbf{y}$ by the Cauchy-Schwartz inequality, it follows from equation (3.7.9) that

$$\mathbf{u}^T H_i \mathbf{u} \geqslant \lambda_i \frac{(\mathbf{d}_i^T \mathbf{u})^2}{\mathbf{g}_i^T H_{i-1} \mathbf{g}_i}, \tag{3.7.10}$$

with equality only if $\mathbf{u}$ is parallel to $\mathbf{y}_i$, whence

$$\mathbf{u}^T H_i \mathbf{u} > 0 \qquad \text{as } \mathbf{y}_i^T \mathbf{d}_i \neq 0,$$

and this proves that H_i is positive-definite.

This algorithm is due originally to Davidon [12], but the definitive presentation is due to Fletcher and Powell [13]. It would seem to be the best general-purpose optimization procedure *making use of derivatives* that is currently available. Fletcher and Reeves argue that their method is superior when storage is restricted, and this may well be so. However, the Davidon method lends itself admirably to implementation on machines with two level stores so that the machine configuration on which the method of Fletcher and Reeves is feasible but that of Davidon is not must be restricted indeed.

Broyden [14] has suggested that the Davidon algorithm might be considered a modification of Newton's method (Chapter 4, Section 8) for solving the system of equations $\partial F/\partial x_i = 0$, $i = 1, 2, \ldots, n$. This is reflected in our presentation of the algorithm (and in equation (3.7.3) in particular). The idea of improving the approximate inverse matrix using information gained in the current descent step is close to the main theme of Broyden's paper. It reappears in the next chapter, in particular in Sections 9 and 12.

The method was called a variable metric method by Davidon. This terminology is misleading because no transformation of the independent variables is attempted. It is based on an analogy between the method of steepest descent and equation (3.7.3), and stems from the observation that when the level surfaces are spheres then the method of steepest descent finds the minimum in one step, while the direction given by equation (3.7.3) finds the minimum in one step when the level surfaces have the equation $\mathbf{x}^T H_i^{-1} \mathbf{x} =$ constant.

3.8. POWELL'S METHOD WITHOUT DERIVATIVES

The methods of the two preceding sections require the explicit calculation of partial derivatives of F. As each of these is, in general, at least as complicated an expression as F this is effectively $(n + 1)$ function values at each step so that a method which avoids the calculation of derivatives has the possibility of being more efficient as well as having the advantage of being much more convenient to use. One such method has been given by Powell [15] and is based on the observation that if the minimum of a positive-definite quadratic form is sought in the direction **v** from each of two distinct points then the vector joining the resulting minima is conjugate to **v**. This is

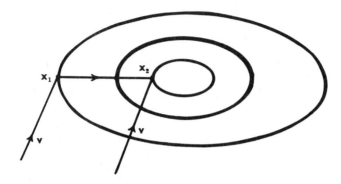

**Figure 3.8.I.   The join of minima found by searching along
v is conjugate to v.**

illustrated in Figure 3.8.1. Let the two minima be $\mathbf{x}_1$ and $\mathbf{x}_2$, respectively; then we have

$$\mathbf{v}^T(A\mathbf{x}_1 + \mathbf{b}) = 0,$$

and

$$\mathbf{v}^T(A\mathbf{x}_2 + \mathbf{b}) = 0,$$

whence, subtracting,

$$\mathbf{v}^T A(\mathbf{x}_1 - \mathbf{x}_2) = 0, \tag{3.8.1}$$

which demonstrates the conjugacy.

Powell's algorithm uses the more elaborate result that if the search for each minimum is made along p conjugate directions then the join of these minima is conjugate to all these directions. This follows from Theorem 3.5.1 because the result of each search depends only on the starting points and not on the order in which each descent step is carried out. Therefore it can be arranged that any of the p conjugate directions can be used last, and this proves the result.

The basic algorithm can now be given. We assume that n independent vectors $\mathbf{d}_1, \ldots, \mathbf{d}_n$ are given initially.

(i) Let λ_0 minimize $F(\mathbf{x}^{(0)} + \lambda \mathbf{d}_n)$, and set

$$\mathbf{x}^{(1)} = \mathbf{x}^{(0)} + \lambda_0 \mathbf{d}_n.$$

(ii) For $i = 1, 2, \ldots, n$ compute λ_i to minimize $F(\mathbf{x}^{(i)} + \lambda \mathbf{d}_i)$ and set

$$\mathbf{x}^{(i+1)} = \mathbf{x}^{(i)} + \lambda_i \mathbf{d}_i.$$

(iii) Set $\mathbf{d}_i = \mathbf{d}_{i+1}$, $i = 1, 2, \ldots, n - 1$.

(iv) Set $\mathbf{d}_n = \mathbf{x}^{(n+1)} - \mathbf{x}^{(i)}$, $\mathbf{x}^{(0)} = \mathbf{x}^{(n+1)}$.

(v) Repeat (i).

This procedure illustrated for $n = 2$, $\mathbf{d}_i = \mathbf{e}_i$, in Figure 3.8.2 (see Kowalik [16]). It will be seen that at the end of the first sweep $\mathbf{x}^{(n+1)} - \mathbf{x}^{(1)}$ is conjugate to $\mathbf{d}_n$, at the end of the second sweep $\mathbf{x}^{(n+1)} - \mathbf{x}^{(1)}$ is conjugate to $\mathbf{d}_n$ and $\mathbf{d}_{n-1}$, and so on. After $n - 1$ complete sweeps n conjugate directions are obtained.

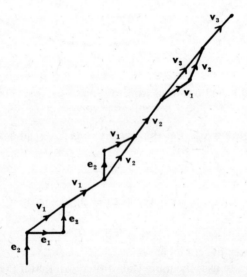

Figure 3.8.2.   Powell's basic algorithm illustrated in two dimensions.

REMARK.   (i) It will be noted that this algorithm can be thought of as a natural development of the Rosenbrock algorithm (Chapter 2, Section 5) in two ways:

(a) The simple searches in each direction are refined to find a minimum rather than just a lower value.

(b) Conjugate directions rather than orthogonal directions are used in these searches.

An algorithm intermediate between those of Rosenbrock and Powell has been given by Davies, Swann, and Campey [17]. In this they define new directions by the relation

$$
Q = D \begin{bmatrix} \lambda_1 & & & \\ \cdot & \lambda_2 & & \\ \cdot & & \cdot & \\ \cdot & & & \cdot \\ \lambda_n & \lambda_n & \cdots & \lambda_n \end{bmatrix}, \tag{3.8.2}
$$

where $\kappa_i(D) = \mathbf{d}_i$, $i = 1, 2, \ldots, n$, and λ_i is the magnitude of the displacement along $\mathbf{d}_i$ in the i-th descent step, and then they orthogonalize the columns of Q to provide the directions for the next cycle of descent steps.

(ii) The Powell algorithm produces a new conjugate direction once in every $(n + 1)$ searches, while the two previous algorithms produces one for every search but require n times as much computation for each function evaluation.

Powell reports that the directions generated by the basic algorithm tend to lose their efficiency as the computation proceeds. Presumably we could restart after a fixed number of iterations as in the algorithm of Fletcher and Reeves; however this does not seem to have been tried in practice. Powell proceeds by using an interesting criterion for estimating whether the direction generated at the current stage is efficient. He rejects it if the criterion fails and computes another cycle of descent steps using the current directions. Otherwise he accepts this new direction, which then replaces one of those used in the current step. Up to a point this involves restarting, but the algorithm is designed both to decide for itself when this is necessary, and to preserve as much useful information as it can from previous stages.

LEMMA 3.8.1. *Let vectors* $\mathbf{x}_1, \ldots, \mathbf{x}_n$ *be scaled so that* $\mathbf{x}_i^T A \mathbf{x}_i = 1$, $i = 1, 2, \ldots, n$. *Let X be the matrix such that* $\kappa_j(X) = \mathbf{x}_j$. *Then* det (X) *is a maximum when the* $\mathbf{x}_j$ *are mutually conjugate with respect to A.*

PROOF. Let the vectors $\mathbf{y}_1, \ldots, \mathbf{y}_n$ be mutually conjugate with respect to A and satisfy the above scaling condition. Then we can write

$$
\mathbf{x}_i = \sum_{j=1}^{n} U_{ij} \mathbf{y}_j
$$
$$
= Y\mathbf{U}_i, \tag{3.8.3}
$$

where Y is the matrix such that $\kappa_j(Y) = \mathbf{y}_j$, and $\mathbf{U}_i$ is the vector with components U_{ij}, $j = 1, 2, \ldots, n$. Let U be the matrix with $\kappa_j(U) = \mathbf{U}_j$, then

$$
X = YU, \tag{3.8.4}
$$

so that
$$\det (X) = \det (Y) \det (U). \tag{3.8.5}$$
Also, by definition
$$1 = \mathbf{x}_i^T A \mathbf{x}_i = \mathbf{U}_i^T Y^T A Y \mathbf{U}_i$$
$$= \mathbf{U}_i^T \mathbf{U}_i, \tag{3.8.6}$$
so that, by Hadamard's lemma,
$$\det (U) \leqslant \prod_{i=1}^{n} (\mathbf{U}_i^T \mathbf{U}_i)^{1/2} = 1, \tag{3.8.7}$$

with equality only if U is an orthogonal matrix (Hadamard's lemma follows by noting that $\det (U)$ has the geometric interpretation as the volume of the parallelipiped generated by the columns of U, and this is a maximum when these are orthogonal when it is equal to the product of their lengths.)

If U is orthogonal then
$$X^T A X = U^T Y^T A Y U = U^T U = I, \tag{3.8.8}$$

so that the vectors $\mathbf{x}_i$ are mutually conjugate, and this completes the proof.

This lemma provides Powell's criterion. He tests the new direction to see if it is possible to add it and drop one of the current set in such a way that the appropriate determinant is not decreased. Of course this must be done in such a way that only function values are used. Let $\boldsymbol{\xi}_i = \mathbf{d}_i/(\mathbf{d}_i^T A \mathbf{d}_i)^{\frac{1}{2}}$, $i = 1$, $2, \ldots, n$, then we write

$$\mathbf{x}^{(n+1)} - \mathbf{x}^{(1)} = \sum_{i=1}^{n} \alpha_i \boldsymbol{\xi}_i, \tag{3.8.9}$$

whence
$$\alpha_i = \boldsymbol{\xi}_i^T A(\mathbf{x}^{(i+1)} - \mathbf{x}^{(i)}), \tag{3.8.10}$$
by conjugacy. Thus

$$\alpha_i^2 = (\mathbf{x}^{(i+1)} - \mathbf{x}^{(i)})^T A(\mathbf{x}^{(i+1)} - \mathbf{x}^{(i)})$$
$$= \mathbf{x}^{(i)T} A \mathbf{x}^{(i)} - \mathbf{x}^{(i+1)T} A \mathbf{x}^{(i+1)} + 2(\mathbf{x}^{(i+1)} - \mathbf{x}^{(i)})^T A \mathbf{x}^{(i+1)}. \tag{3.8.11}$$

But the condition for the minimum in the direction $\boldsymbol{\xi}_i$ is
$$(\mathbf{x}^{(i+1)} - \mathbf{x}^{(i)})^T (A \mathbf{x}^{(i+1)} + \mathbf{b}) = 0,$$
whence
$$(\mathbf{x}^{(i+1)} - \mathbf{x}^{(i)})^T A \mathbf{x}^{(i+1)} = -(\mathbf{x}^{(i+1)} - \mathbf{x}^{(i)})^T \mathbf{b}, \tag{3.8.12}$$
and this gives (on substituting (3.8.12) into equation (3.8.11))

$$\alpha_i^2 = 2\{\tfrac{1}{2}\mathbf{x}^{(i)T} A \mathbf{x}^{(i)} - \tfrac{1}{2}\mathbf{x}^{(i+1)T} A \mathbf{x}^{(i+1)} - (\mathbf{x}^{(i+1)} - \mathbf{x}^{(i)})^T \mathbf{b}\}$$
$$= 2(F(\mathbf{x}^{(i)}) - F(\mathbf{x}^{(i+1)})). \tag{3.8.13}$$

Thus we have proved the result.

LEMMA 3.8.2. *If ξ^i is scaled so that $\xi_i^T A \xi^i = 1$ then the component of* $x^{(n+1)} - x^{(1)}$ *in the direction of* ξ_i *is*

$$\alpha_i = \sqrt{2(F(x^{(i)}) - F(x^{(i+1)}))}.$$

COROLLARY. *If $\eta^T A \eta = 1$, then the minimum in the direction η from the point* y *is at*

$$x = y + \sqrt{2(F(y) - F(x))}\eta. \tag{3.8.14}$$

PROOF. This follows from a slight reorganization of the algebra leading to Lemma 3.8.2.

If we write

$$x^{(n+1)} - x^{(1)} = \mu\xi, \tag{3.8.15}$$

where $\xi^T A \xi = 1$, then

$$\det(\xi_1, \ldots, \xi_{p-1}, \xi, \xi_{p+1}, \ldots, \xi_n) = \frac{\alpha_p}{\mu} \det(\xi_1, \ldots, \xi_n), \tag{3.8.16}$$

so that if we are to add ξ, then we must drop that ξ_p for which

$$\sqrt{2(F(x^{(p)} - F(x^{(p+1)}))}$$

is a maximum (in modulus). This ensures that the new determinant of directions is as large as possible. To compute μ let $x^{(n+1)} + \gamma\xi$ be the minimum in the direction ξ through the point $x^{(n+1)}$. Then by the corollary to Lemma 3.8.2 we have

$$x^{(n+1)} + \gamma\xi = x^{(n+1)} + \sqrt{2(F(x^{(n+1)}) - F_s)}\xi$$

$$= x^{(1)} + \sqrt{2(F(x^{(1)}) - F_s)}\xi,$$

where $F_s = F(x^{(n+1)} + \gamma\xi)$, whence

$$0 = \xi^T A(x^{(n+1)} - x^{(1)}) + \sqrt{2(F(x^{(n+1)}) - F_s)} - \sqrt{2(F(x^{(1)}) - F_s)},$$

so that

$$\mu = \sqrt{2(F(x^{(1)}) - F_s)} - \sqrt{2(F(x^{(n+1)}) - F_s)}. \tag{3.8.17}$$

REMARK. If $x^{(n+1)} + \gamma\xi$ lies between $x^{(1)}$ and $x^{(n+1)}$ then the negative square root must be taken in forming $\sqrt{2(F(x^{(n+1)}) - F_s)}$. In this case

$$\mu = \sqrt{(2F(x^{(1)} - F_s)} + |\sqrt{2(F(x^{(n+1)}) - F_s)}|$$

$$= \sqrt{2(F(x^{(1)}) - F(x^{(n+1)}) + F(x^{(n+1)}) - F_s)} + |\sqrt{2(F(x^{(n+1)}) - F_s)}|,$$

whence

$$|\mu| \geqslant |\sqrt{2(F(\mathbf{x}^{(1)}) - F(\mathbf{x}^{(n+1)}))}| > |\alpha_p|. \qquad (3.8.18)$$

Powell's criterion is that ξ replaces ξ_p provided $|\alpha_p/\mu| \geqslant 1$. In this case we replace $\mathbf{d}_p$ by $\mathbf{x}^{(n+1)} - \mathbf{x}^{(1)}$ and continue the iteration. Otherwise we continue with another descent cycle using $\mathbf{d}_1, \ldots, \mathbf{d}_n$. If the minimum along ξ from $\mathbf{x}^{(n+1)}$ lies between $\mathbf{x}^{(1)}$ and $\mathbf{x}^{(n+1)}$ then this criterion automatically fails by equation (3.8.18).

Fletcher [18] has compared this method with two others that do not require derivatives. He notes that Powell's method gives an efficient rate of ultimate convergence when F can be adequately represented by a quadratic form in the neighborhood of the minimum, and that the method of Davies, Swann, and Campey is competitive when this assumption on F does not hold.

He also notes that Powell's criterion for selecting a new direction might be too rigid for problems involving a large number of variables so that new directions are not chosen as often as might be desirable. It seems that more numerical evidence is desirable, and we return to this point in Chapter 6.

REFERENCES

1. A. M. Ostrowski, *Solution of Equations and Systems of Equations*, 2nd ed., Academic Press, New York and London, 1966.
2. G. E. Forsythe and W. R. Wasow, *Finite-Difference Methods for Partial Differential Equations*, John Wiley & Sons, New York, 1960.
3. R. Varga, *Matrix Iterative Analysis*, Prentice-Hall, Englewood Cliffs, N.J., 1962.
4. E. L. Stiefel, Über einiger Methoden der Relaxationsrechnung, *Z. Angew. Math. Phys.*, *3*(1952), 1.
5. Hirotugu Akaike, On a Successive Transformation of Probability Distribution and Its Application to the Analysis of the Optimum Gradient Method, *Ann. Inst. Statist. Math., Tokyo*, *11*(1959), 1.
6. G. E. Forsythe and T. S. Motzkin, Asymptotic Properties of the Optimum Gradient Method (abstract), *Bull. Amer. Math. Soc.*, *57*(1951), 183.
7. G. E. Forsythe, *On the Asymptotic Directions of the S-Dimensional Optimum Gradient Method*, Computer Science Department, Standford University, Technical Report No. CS61, 1967.
8. R. Fletcher and C. M. Reeves, Function Minimization by Conjugate Gradients, *Computer J.*, *7*(1964), 149.
9. M. R. Hestenes and E. L. Stiefel, Methods of Conjugate Gradients for Solving Linear Systems, *J. Res. Nat. Bur. Standards, Sect. B*, *49*(1952), 409.
10. E. L. Stiefel, Kernel Polynomials in Linear Algebra and Their Numerical Applications, *Nat. Bur. Standards Appl. Math. Ser.*, *49*(1958).
11. F. S. Beckman, The Solution of Linear Equations by the Conjugate Gradient Method, in *Mathematical Methods for Digitial Computers*, Vol. 1, A. Ralston and H. S. Wilf, eds., John Wiley & Sons, New York, 1960.

12. W. C. Davidon, *Variable Metric Method for Minimization*, A.E.C. Research and Development Report, ANL-5990 (rev), 1959.
13. R. Fletcher and M. J. D. Powell, A Rapidly Convergent Descent Method for Minimization, *Computer J.*, *6*(1963), 163.
14. C. G. Broyden, A Class of Methods for Solving Nonlinear Simultaneous Equations, *Math. Comp.*, *19*(1965), 577.
15. M. J. D. Powell, An Efficient Method for Finding the Minimum of a Function of Several Variables without Calculating Derivatives, *Computer J.*, *7*(1964), 155.
16. J. Kowalik, Nonlinear Programming Procedures and Design Optimization, *Acta Polytech. Scandinav.*, *13*(1966)29.
17. W. H. Swann, *Report on the Development of a New Direct Search Method of Optimization*, I.C.I. Ltd., Central Instrument Laboratory Research Note 64/3, 1964.
18. R. Fletcher, Function Minimization without Evaluating Derivatives—A Review, *Computer J.*, *8*(1965), 33.

Chapter 4

LEAST SQUARES PROBLEMS

4.1. INTRODUCTION

Special methods are available for the particular case in which the function to be minimized has the form of a sum of squares, and these methods form the subject matter for this chapter.

It is convenient first to consider the case in which the function to be minimized is a positive-definite quadratic form, as the techniques used for solving this problem are employed also in setting up iterations in the non-linear case. In the linear case the classic Choleski factorization of a symmetric positive-definite matrix has been the main tool employed in practice, and this is described in Section 4. However, recent work by Golub in particular has shown that in certain important cases when the function is either zero or very small at the minimum then an alternative approach based on orthogonal transformations is preferable for numerical work. This work is considered in Sections 5 and 6. In Section 7 an application to regression analysis is considered, and it is shown that this particular problem is conveniently treated by a variant of the classic technique of Jordan elimination.

The treatment of nonlinear problems has several unifying factors. First, the Newton iteration for systems of n equations in n unknowns is a particular case of the Gauss iteration for systems of n equations in $p < n$ unknowns. Second, the secant iteration is presented as an attempt to economize on the amount of calculation involved in the Newton iteration, and it is shown that the secant iteration can be presented in a form that, when generalized in an obvious way to the case $n > p$, gives an algorithm that is closely related to one due to Powell. Finally the orthogonal transformation technique of Golub has its algorithmic possibilities, and certain of these are indicated.

4.2. THE LINEAR LEAST SQUARES PROBLEM

Let A be a matrix of n rows and p columns where $p < n$, and let the rank of A be p. It is in general impossible to solve the system of equations

$$A\mathbf{x} = \mathbf{b}, \tag{4.2.1}$$

for arbitrary right-hand side **b**. However, for any **x** a relation of the form

$$Ax = b + r, \tag{4.2.2}$$

holds. This suggests the possibility of seeking an **x** which minimizes $\|r\|$ in some appropriate norm. Here we concern ourselves with the Euclidean vector norm, and we refer to **x** as the *least squares* solution to (4.2.1).

A typical application is to the curve-fitting problem to determine a_j, $j = 1, 2, \ldots, p$, to minimize $\sum_{i=1}^{n} r_i^2$ where

$$r_i = f(x_i) - \sum_{j=1}^{p} a_j \phi_j(x_i). \tag{4.2.3}$$

A closely related application is regression analysis where, from a series of observations of certain variables characteristic of a system, it is desired to determine if the behavior of certain of the variables depends significantly on some linear combination of the others. The algebraic formulation is identical with that for the curve-fitting problem.

4.3. THE NORMAL EQUATIONS

To minimize $\|r\|$ is a simple exercise in the calculus of several variables. From

$$\|r\|^2 = (Ax - b)^T (Ax - b) \tag{4.3.1}$$

partial differentiation gives the following necessary conditions for a stationary point

$$\frac{\partial}{\partial x_j} \|r\|^2 = 2e_j^T A^T Ax - 2e_j^T A^T b$$

$$= 0, \quad j = 1, 2, \ldots, p. \tag{4.3.2}$$

Thus **x** must satisfy the set of linear equations (the *normal equations*)

$$A^T Ax = A^T b. \tag{4.3.3}$$

As A has rank p, $Ax = 0$ only if $x = 0$. It follows that the matrix $M = A^T A$ is positive-definite and hence is nonsingular. Thus equation (4.3.3) determines **x** uniquely.

4.4. THE CHOLESKI DECOMPOSITION OF A POSITIVE-DEFINITE MATRIX

Equation (4.3.3) is traditionally solved by factorizing the matrix M into the form LL^T where L is lower triangular (the Choleski decomposition of a positive-definite matrix [1]). The feasibility of this factorization is most readily demonstrated by induction.

Let M_i be the leading $i \times i$ principal minor of M. It is clear that M_i is positive-definite. Assume that $M_{i-1} = L_{i-1}L_{i-1}^T$ and write

$$\begin{bmatrix} L_{i-1} & \\ \mathbf{l}_{i-1}^T & c_i \end{bmatrix}\begin{bmatrix} L_{i-1}^T & \mathbf{l}_{i-1} \\ & c_i \end{bmatrix} = M_i = \begin{bmatrix} M_{i-1} & \mathbf{v}_{i-1} \\ \mathbf{v}_{i-1}^T & M_{ii} \end{bmatrix} \qquad (4.4.1)$$

Such an equation is possible if

$$L_{i-1}\mathbf{l}_{i-1} = \mathbf{v}_{i-1}$$

and

$$c_i^2 + \mathbf{l}_{i-1}^T\mathbf{l}_{i-1} = M_{ii}.$$

A forward substitution gives $\mathbf{l}_{i-1}$, and it remains to show that c_i is real—that is, $M_{ii} > \mathbf{l}_{i-1}^T\mathbf{l}_{i-1}$. This is an immediate consequence of the positive definiteness of M_i which gives

$$0 < [\mathbf{a}^T, b]\begin{bmatrix} M_{i-1} & \mathbf{v}_{i-1} \\ \mathbf{v}_{i-1}^T & M_{ii} \end{bmatrix}\begin{bmatrix} \mathbf{a} \\ b \end{bmatrix}$$

$$= \mathbf{a}^T M_{i-1}\mathbf{a} + 2b\mathbf{v}_{i-1}^T\mathbf{a} + b^2 M_{ii}. \qquad (4.4.2)$$

The desired result follows on writing $\mathbf{a} = (L_{i-1}^T)^{-1}\mathbf{l}_{i-1}$, $b = -1$, and this establishes the induction.

From the inequality

$$M_{ii} > \mathbf{l}_{i-1}^T\mathbf{l}_{i-1}$$

it follows in particular that $L_{ij}^2 < M_{ii}$. Thus the numbers that occur in the Choleski decomposition are bounded by the diagonal elements of the original matrix, and the large multipliers that can occur in the LU factorization of a general matrix are not a problem here. In fact the Choleski algorithm has been shown to be extremely stable [2].

However ill-conditioning is often a problem in linear least squares calculations. To explain this we note that the size of the condition number of a square matrix (which is defined to be $\|A\| \ \|A^{-1}\|$) gives a good indication of the sensitivity of A^{-1} to small perturbations in A [3]. Now the condition number of $A^T A$ is equal to the square of the condition number of A, so that, for square matrices at any rate, the forming of $A^T A$ is generally a retrograde step in trying to solve linear equations with matrix A.

It is probable that something similar is true when A is rectangular (this will be made more precise later), so that an algorithm for solving the least squares problem which avoids forming $A^T A$ either explicitly or implicity is of interest.

4.5. FACTORIZATION BY ELEMENTARY ORTHOGONAL MATRICES

Such a method has been given recently by Golub [4]. It is based on a factorization of the rectangular matrix A of the form

$$A = W\begin{bmatrix} U \\ \overline{0} \end{bmatrix}, \qquad (4.5.1)$$

where W is an $n \times n$ orthogonal matrix, and U is a $p \times p$ upper triangular matrix. As W is orthogonal we have

$$A^T A = U^T U,$$

so that $U = L^T$ in the Choleski decomposition.

If this factorization of A is substituted in the expression (4.3.1) for the sum of squares there results

$$\|\mathbf{r}\|^2 = \left(W\begin{bmatrix} U \\ \overline{0} \end{bmatrix}\mathbf{x} - \mathbf{b}\right)^T \left(W\begin{bmatrix} U \\ \overline{0} \end{bmatrix}\mathbf{x} - \mathbf{b}\right)$$

$$= \left(\begin{bmatrix} U \\ \overline{0} \end{bmatrix}\mathbf{x} - W^T\mathbf{b}\right)^T \left(\begin{bmatrix} U \\ \overline{0} \end{bmatrix}\mathbf{x} - W^T\mathbf{b}\right)$$

$$= \begin{bmatrix} U\mathbf{x} - \mathbf{c}_1 \\ \overline{\mathbf{c}_2} \end{bmatrix}^T \begin{bmatrix} U\mathbf{x} - \mathbf{c}_1 \\ \overline{\mathbf{c}_2} \end{bmatrix}, \qquad (4.5.2)$$

where $\begin{bmatrix} \mathbf{c}_1 \\ \overline{\mathbf{c}_2} \end{bmatrix} = W^T\mathbf{b}$.

Equation (4.5.2) shows that the sum of squares has a minimum value of $\mathbf{c}_2^T\mathbf{c}_2$ that is attained when $\mathbf{x}$ satisfies

$$U\mathbf{x} = \mathbf{c}_1. \qquad (4.5.3)$$

To compute the factorization (4.5.1) of A it is convenient to build up the orthogonal matrix W as a product of elementary orthogonal matrices having the form

$$W_i = I - 2\mathbf{w}_i\mathbf{w}_i^T, \quad \mathbf{w}_i^T\mathbf{w}_i = 1. \qquad (4.5.4)$$

Consider now

$$A = W_1 W_1 A = W_1 A_1, \qquad (4.5.5)$$

where $A_1 = W_1 A$. It is possible to choose W_1 so that A_1 has zeros in the first column except in the (1, 1) position. In this case we must have

$$\kappa_1(A_1) = \lambda_1\mathbf{e}_1 = W_1\kappa_1(A), \qquad (4.5.6)$$

whence

$$\lambda_1^2 = \sum_{i=1}^{n} A_{i1}^2, \tag{4.5.7}$$

which follows on taking the scalar product of equation (4.5.6) with itself and remembering that W_1 is orthogonal. Using equation (4.5.4) we see that

$$\mathbf{w}_1 = \gamma_1(\kappa_1(A) - \lambda_1 \mathbf{e}_1), \tag{4.5.8}$$

where γ_1 is to be chosen so that $\|\mathbf{w}_1\| = 1$. This gives

$$\gamma_1 = 1/(2\lambda_1(\lambda_1 - A_{11}))^{1/2}. \tag{4.5.9}$$

As equation (4.5.7) does not fix the sign of λ_1 it can be chosen to avoid cancellation in equation (4.5.9), and this gives

$$\text{sgn}\,(\lambda_1) = -\text{sgn}\,(A_{11}). \tag{4.5.10}$$

In the next stage of the algorithm we set

$$A_1 = W_2 W_2 A_1 = W_2 A_2,$$

where W_2 is chosen so that the second column of A_2 is zero below the $(2, 2)$ element. To avoid destroying the zeros introduced in the first stage of the computation it is necessary that $(\mathbf{w}_2)_1 = 0$, and this choice also leaves the first row of A_1 unaltered. Thus the algebra in this case is identical with that of the first transformation, but it is applied only to the $(n - 1) \times (p - 1)$ matrix formed by deleting the first row and column of A_1.

At this point the method of procedure should be clear, and after at most p transformations $((p - 1)$ if A is square), the form of equation (4.5.1) is obtained with W given by

$$W = \prod_{i=1}^{p} W_i. \tag{4.5.11}$$

4.6. THESE FACTORIZATIONS COMPARED

To compare these algorithms it is necessary to have a measure of the sensitivity of the problem with respect to small perturbations (for example round-off error), and it proves convenient to define the condition number of a rectangular matrix to be [5]

$$\chi(A) = \sigma_1/\sigma_n, \tag{4.6.1}$$

where

$$\sigma_1 = \max \|A\mathbf{x}\|/\|\mathbf{x}\|, \qquad \|\mathbf{x}\| \neq 0,$$

$$\sigma_n = \min \|A\mathbf{x}\|/\|\mathbf{x}\|, \qquad \|\mathbf{x}\| \neq 0.$$

Clearly σ_1^2 and σ_n^2 are the greatest and least eigenvalues of $A^T A$. As $\chi(A)$ is invariant under orthogonal transformation we have $\sigma_1 = \|U\|$, $\sigma_n = 1/\|U^{-1}\|$.

For the Choleski algorithm we have to solve a set of linear equations with matrix $A^T A$. In this case we have

$$\chi(A^T A) = \chi(A)^2, \qquad (4.6.2)$$

so that the condition number for this algorithm depends on the square of the measure we have adopted for the sensitivity of A. It is tempting to assume that the condition number of the algorithm based on the orthogonal factorization would be just $\chi(A) = \|U\| \, \|U^{-1}\|$, but this turns out to be an over-simplification.

To see why this should be we consider carefully the computation of the algorithm allowing at each stage perturbations due to rounding errors. Instead of the exact orthogonal matrix we record a matrix V where

$$V = W + E_1, \qquad (4.6.3)$$

and, similarly, we record an upper triangular matrix $\bar{U}$ where

$$\bar{U} = U + E_2. \qquad (4.6.4)$$

The computed solution will satisfy

$$\mathbf{y} = [\bar{U}^{-1} | 0] V^T \mathbf{b} + \mathbf{f}_1$$
$$= [U^{-1} | 0] W^T \mathbf{b} + [\bar{U}^{-1} - U^{-1} | 0] W^T \mathbf{b} + [\bar{U}^{-1} | 0] E_1^T \mathbf{b} + \mathbf{f}_1, \quad (4.6.5)$$

whence

$$\mathbf{y} - \mathbf{x} = [\bar{U}^{-1} - U^{-1} | 0] W^T \mathbf{b} + [\bar{U}^{-1} | 0] E_1^T \mathbf{b} + \mathbf{f}_1. \qquad (4.6.6)$$

From equation (4.6.4) we have

$$U^{-1} = \bar{U}^{-1} + U^{-1} E_2 \bar{U}^{-1}, \qquad (4.6.7)$$

whence

$$\bar{U}^{-1} - U^{-1} = -U^{-1} E_2 \{I + U^{-1} E_2\}^{-1} U^{-1}. \qquad (4.6.8)$$

Substituting from equation (4.6.8) into equation (4.6.5) gives

$$\mathbf{y} - \mathbf{x} = [-U^{-1} E_2 \{I + U^{-1} E_2\}^{-1} U^{-1} | 0] W^T \mathbf{b} + [U^{-1} | 0] E_1^T \mathbf{b} + \mathbf{f}_1. \quad (4.6.9)$$

Up to this point the analysis is effectively independent of whether A is rectangular or square. To introduce this distinction we write

$$\mathbf{b} = W \begin{bmatrix} U \\ 0 \end{bmatrix} \mathbf{x} - \mathbf{r}, \qquad (4.6.10)$$

so that $W^T\mathbf{b}$ can be written

$$W^T\mathbf{b} = \left[\frac{U}{0}\right]\mathbf{x} + \left[\frac{\mathbf{s}_1}{\mathbf{s}_2}\right] \tag{4.6.11}$$

If this equation is substituted in equation (4.6.9) there results

$$\mathbf{x} - \mathbf{y} = [U^{-1}E_2\{I + U^{-1}E_2\}^{-1}](\mathbf{x} + U^{-1}\mathbf{s}_1) - [\bar{U}^{-1} \,|\, 0]E_1^T\mathbf{b} - \mathbf{f}_1. \tag{4.6.12}$$

The critical term is that involving $\mathbf{s}_1$. This is absent if A is square, or if A is rectangular and $\mathbf{r} = 0$ (so that the equations are consistent), and if it is absent then $\|\mathbf{x} - \mathbf{y}\|$ can be estimated by an expression involving $\|U^{-1}\|$ but not any higher power. If $\mathbf{s}_1 \neq 0$, then the estimate for $\|\mathbf{x} - \mathbf{y}\|$ must contain a term $\|U^{-1}\|^2$. It is no restriction to assume $\|A\| = \|U\| = 1$ as, in practice, A would be scaled so that $\|A\|$ is close to 1 so that we can identify $\|U^{-1}\|$ with $\chi(A)$.

We have demonstrated that in any application in which $\|\mathbf{r}\|$ is either zero or very small the orthogonal factorization has a sensitivity to perturbations proportional to $\chi(A)$ while the Choleski decomposition has a sensitivity proportional to $\chi(A)^2$. If $\|\mathbf{r}\|$ is not small compared with $\chi(A)$ then the sensitivity of both algorithms depends on $\chi(A)^2$.

4.7. AN ALGORITHM FOR REGRESSION ANALYSIS

A regression analysis can make use of either of the previous algorithms to test the hypothesis that the behavior of a certain variable is predicted by a linear combination of certain other variables. However such a procedure could not serve for two important applications: (i) to determine a minimum set of variables to describe the behavior of a selected variable, and (ii) to suggest dependencies among the variables characterizing a system.

An important first step toward providing an algorithm for these problems is to find a procedure for solving equation (4.3.3) each step of which can be interpreted as solving a regression with fewer variables (or partial regression). Now the normal equations for the partial regression in which the dependent variable is regressed on the first k independent variables can be obtained from equation (4.3.1) by setting $x_{k+1} = x_{k+2} = \cdots = x_p = 0$; and it is readily verified that the matrix of the normal equations in this case is the leading $k \times k$ minor of equation (4.3.3). Thus any method of solving this equation which after k steps transforms the minor into the $k \times k$ unit matrix also gives the solution of the k-th partial regression.

An algorithm to achieve this can be based on transformations by elementary matrices after the manner of the algorithm based on elementary orthogonal transformations. Here the appropriate matrices are the elementary Jordan matrices,

$$J_i = I - \mathbf{j}_i \mathbf{e}_i^T. \tag{4.7.1}$$

In this case we write equation (4.3.3) in the form

$$J_1 M \mathbf{x} = J_1 A^T \mathbf{b}, \tag{4.7.2}$$

and $\mathbf{j}_1$ is chosen so that $\kappa_1(J_1 M)$ satisfies

$$\kappa_1(J_1 M) = \kappa_1(M) - M_{11}\mathbf{j}_1 = \mathbf{e}_1, \tag{4.7.3}$$

whence

$$\mathbf{j}_1 = \frac{1}{M_{11}}(\kappa_1(M) - \mathbf{e}_1). \tag{4.7.4}$$

Note that J_2 cannot alter the first column of $J_1 M$ for

$$J_2 \kappa_1(J_1 M) = \mathbf{e}_1 - \mathbf{j}_2 \mathbf{e}_2^T \mathbf{e}_1 = \mathbf{e}_1, \tag{4.7.5}$$

so we choose J_2 so that

$$J_2 \kappa_2(J_1 M) = \mathbf{e}_2. \tag{4.7.6}$$

Clearly the process can be repeated, and after k stages we have

$$J_k J_{k-1} \cdots J_1 M \mathbf{x} = J_k J_{k-1} \cdots J_1 A^T \mathbf{b}. \tag{4.7.7}$$

The leading $k \times k$ minor has been reduced to the $k \times k$ unit matrix so that the first k components of the right-hand side give the solution to the k-th partial regression. There is no reason why the variables chosen should be the first k in order, and in principle any k could have been chosen.

Ideally the variable to be brought into the regression should be the one that gives the smallest value of $\|\mathbf{r}\|$ at the end of this stage. This cannot be done in general without computing the next partial regression for each of the remaining variables in turn. However a satisfactory strategy can be based on the following argument. Let the variables in the regression at the t-th stage be $x_{\sigma_i}^{(t)}$, $i = 1, \ldots, t$; then the residual at this stage is given by

$$\mathbf{r}^{(t)} = \mathbf{b} - \sum_{i=1}^{t} x_{\sigma_i}^{(t)} \kappa_{\sigma_i}(A). \tag{4.7.8}$$

We can ask which of the remaining variables best predicts $\mathbf{r}^{(t)}$. In other words we want to minimize $\|\mathbf{r}_s^{(t)}\|$ where

$$\mathbf{r}_s^{(t)} = \mathbf{r}^{(t)} - x\kappa_s(A), \tag{4.7.9}$$

where s ranges over the remaining columns of A. It is readily verified that the appropriate value of s is that which minimizes $(\mathbf{r}^{(t)T}\kappa_s(A))^2/M_{ss}$. If this maximum occurs for $s = m$, then we have

$$\|\mathbf{r}^{(t+1)}\| \leqslant \|\mathbf{r}_m^{(t)}\| = \left(\|\mathbf{r}^{(t)}\|^2 - \frac{(\mathbf{r}^{(t)T}\kappa_m(A))^2}{M_{mm}}\right)^{1/2}, \qquad (4.7.10)$$

and this inequality provides a justification for this strategy.

However it may be that a variable in the regression may not be contributing significantly—in other words $\|\mathbf{r}^{(t)}\|$ is not changed appreciably by removing the variable—so that it is convenient to remove it at this stage. This means that we require an algorithm which permits us to exchange variables to and from the regression at each stage. An algorithm permitting this seems first to have been given by Efroymson [6].

To see how this exchange of variables can be organized consider the equation

$$I\mathbf{y} + M\mathbf{x} = A^T\mathbf{b} = \mathbf{c}. \qquad (4.7.11)$$

This reduces to the normal equations when $\mathbf{y} = 0$. Now consider the i-th row of this equation. If this is solved for x_k we have

$$x_k = \frac{1}{M_{ik}}\left\{c_i - y_i - \sum_{j \neq i} M_{ij}x_j\right\}. \qquad (4.7.12)$$

If this equation is substituted back into equation (4.7.11) there results

$$I\begin{bmatrix} y_1 \\ \cdot \\ \cdot \\ y_{i-1} \\ x_k \\ \cdot \\ \cdot \\ y_p \end{bmatrix} + \bar{M}\begin{bmatrix} x_1 \\ \cdot \\ \cdot \\ x_{k-1} \\ y_i \\ \cdot \\ \cdot \\ x_p \end{bmatrix} = \bar{\mathbf{c}}, \qquad (4.7.13)$$

where

$$\bar{M}_{st} = M_{st} - \frac{M_{sk}M_{it}}{M_{ik}} \qquad \begin{cases} s \neq i \\ t \neq k \end{cases}$$

$$= -M_{sk}/M_{ik} \qquad \begin{cases} s \neq i \\ t = k \end{cases}$$

$$= -M_{it}/M_{ik} \qquad \begin{cases} s = i \\ t \neq k \end{cases}$$

$$= -1/M_{ik} \qquad \begin{cases} s = i \\ t = k, \end{cases}$$

and where $\bar{\mathbf{c}}$ is defined in similar fashion. After precisely p steps all components of $\mathbf{x}$ and $\mathbf{y}$ will be exchanged, and the solution of the original set of equations can be obtained by setting $\mathbf{y} = 0$. After k exchanges the solution of the k-th partial regression is obtained by setting to zero $\mathbf{y}$ and $x_{k+1}, \ldots, x_p$. At this point the connection with the previous algorithm based on Jordan elimination should be clear. However this second formulation shows how to remove variables from the regression. All that is required is to reexchange the appropriate x_s with any one of the y_i exchanged at a previous step.

4.8. NEWTON'S METHOD FOR SYSTEMS OF NONLINEAR EQUATIONS

The solution of systems of nonlinear equations in several variables can be obtained by using techniques for solving optimization problems. For example, the problem of solving the system of n equations in n unknowns

$$\mathbf{f}(\mathbf{x}) = 0, \qquad (4.8.1)$$

is equivalent to the problem of minimizing $\phi(\mathbf{f})$ where ϕ is any nonnegative function such that $\phi(0) = 0$ in the sense that the minima of ϕ include the solutions to equation (4.8.1). An obvious choice for ϕ is

$$\phi(\mathbf{f}) = \|\mathbf{f}\|. \qquad (4.8.2)$$

The majority of techniques for solving nonlinear equations have been developed for the special case of a single equation [7]. In deriving these techniques considerable attention has been paid to developing methods that have a fast rate of ultimate convergence, and the development of methods with good global convergence characteristics has been somewhat neglected. However, in line with the approach of the previous chapter, if the correction generated in the current step of the iteration is used only to define a direction, the minimum of $\phi(\mathbf{f})$ sought in this direction, and this minimum is used to define the next approximation, then the global convergence properties of the resulting method should be good and the ultimate rate of convergence little impaired.

Consider, for example, the standard Newton iteration. Let $\mathbf{x}^{(i)}$ be an approximation to the solution to equation (4.8.1), then

$$\mathbf{f}(\mathbf{x}^{(i)} + \mathbf{h}) = 0. \qquad (4.8.3)$$

The left-hand side of this equation is expanded in a Taylor series, and a correction $\mathbf{h}^{(i)}$ is computed by ignoring second- and higher-order terms in the components of **h**. Thus $\mathbf{h}^{(i)}$ satisfies the set of linear equations

$$\mathbf{f}(\mathbf{x}^{(i)}) + J^{(i)}\mathbf{h}^{(i)} = 0, \tag{4.8.4}$$

where

$$J_{st}^{(i)} = \frac{\partial f_s}{\partial x_t}(\mathbf{x}^{(i)}).$$

This matrix will be called the Jacobian. It will always be assumed to be nonsingular.

If the Taylor series is written down to terms of second order using the mean value theorem for functions of several variables then we have

$$\mathbf{f}(\mathbf{x}^{(i)} + \mathbf{h}) = \mathbf{f}(\mathbf{x}^{(i)}) + J^{(i)}\mathbf{h} + \sum_{s,t} \overline{\frac{\partial^2 \mathbf{f}}{\partial x_s \, \partial x_t}} h_s h_t, \tag{4.8.5}$$

where the bar indicates that mean values of the arguments of the components are appropriate. Now $\mathbf{h}^{(i+1)}$ satisfies

$$\mathbf{f}(\mathbf{x}^{(i)} + \mathbf{h}^{(i)}) + J^{(i+1)}\mathbf{h}^{(i+1)} = 0, \tag{4.8.6}$$

so that, using equations (4.8.4) and (4.8.5),

$$\sum_{s,t} \overline{\frac{\partial^2 \mathbf{f}}{\partial x_s \, \partial x_t}} h_s^{(i)} h_t^{(i)} + J^{(i+1)}\mathbf{h}^{(i+1)} = 0, \tag{4.8.7}$$

whence

$$\|\mathbf{h}^{(i+1)}\| \leqslant K \|\mathbf{h}^{(i)}\|^2, \tag{4.8.8}$$

provided suitable assumptions are made on **f**.

An iteration satisfying an inequality of the form (4.8.8) is said to have *second-order* convergence. This should be distinguished from the *quadratic convergence* of the conjugate gradient methods of the previous chapter.

We call the iteration in which we search for a minimum in the direction defined by $\mathbf{h}^{(i)}$ to determine the next approximation $\mathbf{x}^{(i+1)}$ the *modified Newton iteration*. We summarize it as follows:

(i) Compute $\mathbf{h}^{(i)}$ from $\mathbf{f}(\mathbf{x}^{(i)}) + J^{(i)}\mathbf{h}^{(i)} = 0$.
(ii) Define $\mathbf{t}^{(i)} = \mathbf{h}^{(i)}/\|\mathbf{h}^{(i)}\|$.
(iii) Let λ^* minimize $g^{(i)}(\lambda) = \phi(\mathbf{x}^{(i)} + \lambda\mathbf{t}^{(i)}))$.
(iv) Let $\mathbf{x}^{(i+1)} = \mathbf{x}^{(i)} + \lambda^*\mathbf{t}^{(i)}$.

The cost of computing each step of this iteration is dominated by two major items:

(i) The n^2 evaluations necessary (in general) to compute the elements of $J^{(i)}$.

(ii) The requirement to solve the set of equations (4.8.4) at each stage. This involves of the order of $n^3/3$ multiplications.

There are thus considerable advantages to be gained if it is possible to effect economies in either of these items. The most obvious way in which this might be achieved is to proceed for several steps at a time keeping J fixed, and it is tempting to offer the following justifications for this tactic.

(i) When we are some distance from the solution, then equation (4.8.4) is used solely to generate a direction for a linear search.

(ii) In the neighborhood of the solution, J is changing only slightly so that, eventually, not changing J is an obvious tactic.

A further justification is that it has proved successful in practice [8].

However the modified Newton iteration has two advantages that might make it preferable at least in difficult cases:

(I) The direction given by equation (4.8.4) is downhill for the minimization of $F = \|\mathbf{f}\|^2$.

PROOF.   Partial differentiation gives

$$(\nabla F)_q = 2 \sum_i f_i \frac{\partial f_i}{\partial x_q},$$

whence

$$\nabla F = 2J^T \mathbf{f}, \tag{4.8.9}$$

so that (using equation (4.7.4))

$$\mathbf{h}^T \nabla F = -2\mathbf{f}^T (J^{-1})^T J^T \mathbf{f}$$
$$= -2F < 0. \tag{4.8.10}$$

REMARK.   (i) This result should strengthen confidence in the global convergence properties of the modified Newton iteration.

(ii) It should be noted, under the circumstances in which a minimum of a function $G(\mathbf{x})$ is to be found by solving the system of equations $f_i(\mathbf{x}) = \partial G/\partial x_i = 0$, that the direction $\mathbf{h}^{(i)}$ is downhill for minimizing $\|\mathbf{f}\|^2$ but it is not necessarily downhill for minimizing G. This is readily seen by considering

$$G = x_1^4 + x_1 x_2 + (1 + x_2)^2$$

at the point $\mathbf{x} = 0$.

(II) If J is fixed for the final stages of the modified Newton iteration then the convergence is at best first order.

PROOF.   In this case equation (4.8.7) becomes

$$(J - J^{(i)})\mathbf{h}^{(i)} + \sum_{s,t} \overline{\frac{\partial^2 f}{\partial x_s\, \partial x_t}}\, h_s^{(i)} h_t^{(i)} + J^{(i+1)}\mathbf{h}^{(i+1)} = 0, \qquad (4.8.11)$$

and the stated result follows immediately from this equation. Here J is the matrix being held constant.

4.9. THE SECANT ALGORITHM

An alternative way to cut down work in the Newton algorithm is to use differences rather than derivatives in computing J. For the special case of one variable this gives the secant algorithm, and this generalizes to more general systems [9]. Following Barnes, let $J^{(i)}$ be the current approximation to the Jacobian. Then his algorithm is as follows:

  (i) Compute $\mathbf{h}^{(i)}$ from $\mathbf{f}(\mathbf{x}^{(i)}) + J^{(i)}\mathbf{h}^{(i)} = 0$.
  (ii) Set $\mathbf{x}^{(i+1)} = \mathbf{x}^{(i)} + \mathbf{h}^{(i)}$,
  (iii) and $J^{(i+1)} = J^{(i)} + D^{(i)}$.

The crux of the algorithm is the calculation of $D^{(i)}$, which we are to consider as a correction to the current approximation to the Jacobian. In the case of a single variable the secant algorithm is obtained by replacing the tangent by the chord in the Newton algorithm—a replacement that is correct for linear systems. In this case the condition of linearity must be that $D^{(i)}$ is chosen to satisfy

$$\mathbf{f}(\mathbf{x}^{(i+1)}) = \mathbf{f}(\mathbf{x}^{(i)}) + (J^{(i)} + D^{(i)})\mathbf{h}^{(i)}. \qquad (4.9.1)$$

There is considerable freedom in the choice of $D^{(i)}$ in this formulation of the algorithm, and if $D^{(i)}$ is taken as a dyadic (perhaps the simplest choice) then

$$D^{(i)} = \frac{\mathbf{f}(\mathbf{x}^{(i+1)})\mathbf{z}^{(i)T}}{\mathbf{z}^{(i)T}\mathbf{h}^{(i)}}, \qquad (4.9.2)$$

where $\mathbf{z}^{(i)}$ is an arbitrary vector. It turns out to be convenient to choose $\mathbf{z}^{(i)}$ to be orthogonal to $\mathbf{h}^{(i-1)}, \dots, \mathbf{h}^{(i-n+1)}$. With this choice we have for $k \leqslant n$

$$\begin{aligned}
J^{(i+k)}\mathbf{h}^{(i)} &= (J^{(i)} + D^{(i)} + \cdots + D^{(i+k-1)})\mathbf{h}^{(i)} \\
&= (J^{(i)} + D^{(i)})\mathbf{h}^{(i)} \\
&= \mathbf{f}(\mathbf{x}^{(i+1)}) - \mathbf{f}(\mathbf{x}^{(i)}).
\end{aligned} \qquad (4.9.3)$$

Barnes, in his paper, also discusses the relationship between his form of the algorithm and the earlier versions, its rate of convergence, and its property of giving the solution to a system of linear equations in a finite number of steps.

This presentation of the algorithm has the unsatisfactory feature that certain of the choices made appear artificial. An alternative form will now be given which has the advantage of being free from apparently arbitrary choices, and which lends itself to implementation. In this version we keep a matrix of directions $T^{(i)}$, and we work not with the Jacobian but with a matrix $G^{(i)}$ given by

$$G^{(i)} = J^{(i)}T^{(i)}. \qquad (4.9.4)$$

It will be seen that

$$G_{pq}^{(i)} \approx \frac{df_p}{dt_q}$$

where $\mathbf{t}_q = \kappa_q(T^{(i)})$. For any vector $\mathbf{v}$ we have

$$\mathbf{v} = \sum_{j=1}^{n} a_j \mathbf{t}_j = T^{(i)}\mathbf{a},$$

so that

$$J^{(i)}\mathbf{v} = J^{(i)}T^{(i)}\mathbf{a} = G^{(i)}\mathbf{a}. \qquad (4.9.5)$$

The algorithm is as follows. We set $\mathbf{f}^{(i)} = \mathbf{f}(\mathbf{x}^{(i)})$.

(i) Compute $\mathbf{h}^{(i)} = -T^{(i)}G^{(i)-1}\mathbf{f}^{(i)}$.

(ii) Then $\mathbf{x}^{(i+1)} = \mathbf{x}^{(i)} + \mathbf{h}^{(i)}$.

(iii) Replace $\mathbf{t}_s$ by $\mathbf{h}^{(i)}/\|\mathbf{h}^{(i)}\|$ so that

$$T^{(i+1)} = T^{(i)} + (\mathbf{h}^{(i)}/\|\mathbf{h}^{(i)}\| - \mathbf{t}_s)\mathbf{e}_s^T.$$

(iv) Replace $\kappa_s(G^{(i)})$ by $(\mathbf{f}^{(i-1)} - \mathbf{f}^{(i)})/\|\mathbf{h}^{(i)}\|$ so that

$$G^{(i+1)} = G^{(i)} + ((\mathbf{f}^{(i+1)} - \mathbf{f}^{(i)})/\|\mathbf{h}^{(i)}\| - \kappa_s(G^{(i)}))\mathbf{e}_s^T.$$

NOTE. (1) The fourth step is correct for linear systems. It is this step that corresponds to choosing $D^{(i)}$ to satisfy equation (4.9.1).

(2) Only $T^{(i)}$ and $G^{(i)-1}$ are required for each stage of the algorithm. As only one column of $G^{(i)}$ is altered in forming $G^{(i+1)}$ it is possible to update $G^{(i)-1}$ to give $G^{(i+1)-1}$, and this makes possible a considerable economy. We have

$$G^{(i+1)} = G^{(i)}\{I + \mathbf{x}\mathbf{e}_s^T\}, \qquad (4.9.6)$$

where

$$\mathbf{x} = G^{(i)-1}\{(\mathbf{f}^{(i+1)} - \mathbf{f}^{(i)})/\|\mathbf{h}^{(i)}\| - \kappa_s(G^{(i)})\}$$

$$= G^{(i)-1}(\mathbf{f}^{(i+1)} - \mathbf{f}^{(i)})/\|\mathbf{h}^{(i)}\| - \mathbf{e}_s,$$

whence (as is readily verified)

$$G^{(i+1)-1} = \left\{ I - \frac{1}{1 - x_s} \mathbf{x} \mathbf{e}_s^T \right\} G^{(i)-1}. \tag{4.9.7}$$

(3) The choice of s determines the column of $T^{(i)}$ to be replaced. In the Barnes algorithm columns are replaced in sequence. However this is not necessary, and the choice should be made to maximize the performance of the algorithm.

We consider next the equivalence of this algorithm with that of Barnes. In this case $J^{(i+1)}$ is determined by

$$J^{(i+1)}\{ T^{(i)} + (\mathbf{h}^{(i)}/\|\mathbf{h}^{(i)}\| - \mathbf{h}^*/\|\mathbf{h}^*\|)\mathbf{e}_s^T \}$$
$$= G^{(i)} + (\Delta \mathbf{f}^{(i)}/\|\mathbf{h}^{(i)}\| - \Delta \mathbf{f}^*/\|\mathbf{h}^*\|)\mathbf{e}_s^T, \tag{4.9.8}$$

where the star is used to indicate the previous value, and Δ is the forward difference operator (so that $\Delta \mathbf{f}^{(i)} = \mathbf{f}^{(i+1)} - \mathbf{f}^{(i)}$). As $J^{(i)}T^{(i)} = G^{(i)}$, we have

$$(J^{(i+1)} - J^{(i)})T^{(i)} + J^{(i+1)}(\mathbf{h}^{(i)}/\|\mathbf{h}^{(i)}\| - \mathbf{h}^*/\|\mathbf{h}^*\|)\mathbf{e}_s^T$$
$$= (\Delta \mathbf{f}^{(i)}/\|\mathbf{h}^{(i)}\| - \Delta \mathbf{f}^*/\|\mathbf{h}^*\|)\mathbf{e}_s^T, \tag{4.9.9}$$

whence, using $J^{(i)}\mathbf{h}^{(i)} = -\mathbf{f}^{(i)}$,

$$(J^{(i+1)} - J^{(i)})\{ I + (\mathbf{h}^{(i)}/\|\mathbf{h}^{(i)}\| - \mathbf{h}^*/\|\mathbf{h}^*\|)\rho_s(T^{(i)-1}) \}$$
$$= \{ \mathbf{f}^{(i+1)}/\|\mathbf{h}^{(i)}\| - (\Delta \mathbf{f}^* - J^{(i)}\mathbf{h}^*)/\|\mathbf{h}^*\| \}\rho_s(T^{(i)-1}). \tag{4.9.10}$$

Now, provided equation (4.9.3) holds, this equation reduces to

$$J^{(i+1)} = J^{(i)} + \mathbf{f}^{(i+1)}\rho_s(T^{(i+1)-1})/\|\mathbf{h}^{(i)}\|. \tag{4.9.11}$$

But equation (4.9.11) is identical with equation (4.9.2) as the rows of $T^{(i+1)-1}$ form a set of vectors orthogonal to the columns of $T^{(i+1)}$, and we have

$$\|\mathbf{h}^{(i)}\| = \|\mathbf{h}^{(i)}\| \, \rho_s(T^{(i+1)-1})\kappa_s(T^{(i+1)})$$
$$= \rho_s(T^{(i+1)-1})\mathbf{h}^{(i)}.$$

Thus the two algorithms are identical provided the second is started in such a fashion that equation (4.9.3) holds initially. This will be so, for example, if a single cycle of a relaxation computation (Chapter 3, Section 3) is performed initially. In this case the starting matrices satisfy $J^{(1)} = G^{(1)}$, and $T^{(1)} = I$. It is interesting to note that considerations of stability led Barnes to recommend an equation similar to equation (4.9.10) rather than equation (4.9.2) for implementing his algorithm.

Note that, in practice, both these algorithms would be modified to include a search for the minimum along the direction $\mathbf{h}^{(i)}$ in the manner already described for the Newton iteration.

4.10. THE GAUSS ALGORITHM FOR OVER-DETERMINED SYSTEMS

The solution of over-determined systems of nonlinear equations raises further problems. In this case the equations are in general not consistent so that a solution must be sought in the sense of least squares. As before, this amounts to finding a vector $\mathbf{x}$ with p components such that $\|\mathbf{r}\|$ is minimized where

$$\mathbf{f}(\mathbf{x}) = \mathbf{r}. \tag{4.10.1}$$

It is assumed that $\mathbf{f}$ has n components and that $n > p$.

Here we adopt an approach similar to that used in discussing Newton's method in Section 8. Thus we expand $\mathbf{f}(\mathbf{x})$ in a Taylor series about the current approximation $\mathbf{x}^{(i)}$ to the solution, and we ignore terms which are of second or higher order in the correction to $\mathbf{x}^{(i)}$. In this case the result is an over-determined system of linear equations, and the solution of this system is sought in the sense of least squares. This system of equations is written

$$\mathbf{f}(\mathbf{x}^{(i)}) + A^{(i)}\mathbf{h}^{(i)} = \mathbf{r}^{(i)}, \tag{4.10.2}$$

where

$$A_{st}^{(i)} = \frac{\partial f_s}{\partial x_t}(\mathbf{x}^{(i)}).$$

The basic algorithm is as follows:

(i) Compute $A^{(i)}$.
(ii) Form $M^{(i)} = A^{(i)T}A^{(i)}$.
(iii) Compute $\mathbf{h}^{(i)}$ from $M^{(i)}\mathbf{h}^{(i)} = -A^{(i)T}\mathbf{f}^{(i)}$.
(iv) Set $\mathbf{x}^{(i+1)} = \mathbf{x}^{(i)} + \mathbf{h}^{(i)}$.

To examine the rate of convergence of this iteration we expand $\mathbf{f}(\mathbf{x}^{(i+1)}) = \mathbf{f}(\mathbf{x}^{(i)} + \mathbf{h}^{(i)})$ in a Taylor series. This gives

$$\mathbf{f}^{(i+1)} = \mathbf{f}^{(i)} + A^{(i)}\mathbf{h}^{(i)} + \mathbf{L}\mathbf{h}^{(i)}, \tag{4.10.3}$$

where $\mathbf{L}^{(i)}$ is an expression for the remainder and $\|\mathbf{L}^{(i)}\| = 0\ (\|\mathbf{h}^{(i)}\|^2)$ by equation (4.8.5). This equation and equation (4.10.2) give

$$A^{(i+1)}\mathbf{h}^{(i+1)} = \mathbf{r}^{(i+1)} - \mathbf{L}^{(i)} - \mathbf{r}^{(i)}, \tag{4.10.4}$$

and the corresponding normal equations are

$$M^{(i+1)}\mathbf{h}^{(i+1)} = -A^{(i+1)T}\mathbf{L}^{(i)} - (A^{(i+1)} - A^{(i)})^T\mathbf{r}^{(i)}$$

$$= -A^{(i+1)T}\mathbf{L}^{(i)} - \|\mathbf{h}^{(i)}\| \frac{d\bar{A}^T}{dt} \mathbf{r}^{(i)}, \tag{4.10.5}$$

where the bar denotes that mean values of the arguments are appropriate. We have written $\mathbf{h}^{(i)} = \|\mathbf{h}^{(i)}\| \mathbf{t}$, and we have used that $A^{(i)T}\mathbf{r}^{(i)} = 0$. This equation shows that $\|\mathbf{h}^{(i+1)}\| = 0(\|\mathbf{h}^{(i)}\|^2)$ only if $\|\mathbf{r}^{(i)}\| = 0(\|\mathbf{h}^{(i)}\|)$. This is possible only if the system (4.10.1) is consistent so that $\|\mathbf{r}\| = 0$ at the solution. In all other cases the iteration is at best first order. In fact the iteration is potentially divergent, and a sufficient condition for divergence from a starting point which is an arbitrary small displacement from the solution is that

$$\left\| M^{-1} \frac{dA^T}{ds} \mathbf{r} \right\| > 1 \tag{4.10.6}$$

for all directions $\mathbf{s}$ through the solution to equation (4.10.1). It is assumed that all quantities in equation (4.10.6) are evaluated at the solution to equation (4.10.1). Equation (4.10.6) will always be satisfied unless $\|\mathbf{r}\|$ is small enough at the solution.

REMARK. This has particular relevance when the minimization problem occurs in estimating parameters by fitting to experimental data for $\|\mathbf{r}\|$ and will be large at the solution when there is considerable error in some or all of the observations [10].

As usual we consider the possibility of employing a modified iteration in which the correction $\mathbf{h}^{(i)}$ is used to define a direction and the next approximation to the solution is sought as the minimum of $\|\mathbf{f}\|^2$ in this direction. In view of the shortcomings of the iteration indicated in the preceding paragraph the following result is particularly important. It generalizes the corresponding result for Newton's method:

(I) The direction $\mathbf{h}^{(i)}$ is "downhill" for the minimization of $F = \|\mathbf{f}\|^2$.

PROOF. We have

$$\nabla F^{(i)} = 2A^{(i)T}\mathbf{f}^{(i)}, \tag{4.10.7}$$

so that

$$\mathbf{h}^{(i)T}\nabla F^{(i)} = -2\mathbf{f}^{(i)T}A^{(i)}M^{(i)-1}A^{(i)T}\mathbf{f}^{(i)}, \tag{4.10.8}$$

whence $\mathbf{h}^{(i)T}\nabla F^{(i)} < 0$ as $M^{(i)}$ is positive-definite.

REMARK. By Ostrowski's theorem (Chapter 3, Section 2) this result shows that the modified iteration converges even when the unmodified iteration is divergent for starting points arbitrarily close to the solution.

It is possible to provide an algorithm with second-order convergence. To do this F is expanded in a Taylor series up to terms of second order in the components of $\mathbf{h}$. The correction to the current approximation is then found by minimizing this quadratic form. It may be noted that this method is quadratically convergent, so that here quadratic and second-order convergence both obtain. Partial differentiation gives

$$\frac{\partial F}{\partial h_k} = 2Q_k + 0(\|\mathbf{h}\|^2), \tag{4.10.9}$$

where

$$Q_k = \sum_i \left\{ f_i \frac{\partial f_i}{\partial x_k} + \sum_j \left(\frac{\partial f_i}{\partial x_k} \frac{\partial f_i}{\partial x_j} + f_i \frac{\partial^2 f_i}{\partial x_k \partial x_j} \right) h_j \right\}, \tag{4.10.10}$$

so that the minimum of the quadratic form satisfies the linear equations

$$Q_k = 0, \qquad k = 1, 2, \ldots, p. \tag{4.10.11}$$

The right-hand side of these equations is $-\nabla F/2$, but the matrix differs from that of the previous algorithm by the addition of a term

$$\sum_i f_i \frac{\partial^2 f_i}{\partial x_j \partial x_k}.$$

That the iteration is second order follows at once from equations (4.10.9) and (4.10.11) for they give

$$\|\nabla F(\mathbf{x}^{(i+1)})\| = 0(\|\mathbf{h}^{(i)}\|^2). \tag{4.10.12}$$

The requirement to evaluate the second derivative terms is a considerable disadvantage of this algorithm, and it appears to have been little used. However the previous algorithm seems to have found fairly wide acceptance despite the problems involved in its use. It is usually called the *Gauss algorithm*, and when varied to permit the search for the minimum of F in the direction $\mathbf{h}^{(i)}$ it is called the *modified Gauss algorithm*. A variant whose use is to be discouraged always takes a correction of the form $\gamma \mathbf{h}^{(i)}$ for some fixed constant γ. It is unlikely that any choice of γ which differs much from 1 could give a satisfactory ultimate rate of convergence when $\|\mathbf{f}\|$ is small at the minimum, while a much smaller value of γ is often required to give convergence from difficult starting points. The point is that, as a substantial amount of work must be done at each stage to generate $A^T A$ and to solve the resulting equations, the extra work required for the one-dimensional minimization is negligible, and the improvement in performance considerable.

4.11. THE APPROACH OF LEVENBERG, MARQUARDT, AND MORRISON

We have seen that the Gauss method is potentially divergent for starting points arbitrarily close to the desired solution unless the system of equations is consistent, but that the direction of the correction generated by the Gauss method is downhill so that the modified Gauss method is (in theory) always convergent. In practice the directions generated may not be very favorable, and Marquardt [11] has reported that over a number of trials he found that the directions generated made a consistently large angle ($> 80°$) with the direction of steepest descent. For this reason he suggests that the correction in the Gauss method be computed from the equation

$$(M + \mu I)\mathbf{h} = -A^T\mathbf{f} = -\mathbf{g}, \qquad (4.11.1)$$

for some $\mu > 0$. Unfortunately there is little information on how to choose μ. However the method has been widely used with considerable success. It was suggested also by Morrison [12] and Levenberg [13].

The results that follow explain the success of the method at least in part:

(I) As μ increases from zero $\|\mathbf{h}(\mu)\|$ decreases monotonically.

PROOF. We have

$$\frac{d}{d\mu}\|\mathbf{h}\| = \frac{d}{d\mu}(\mathbf{h}^T\mathbf{h})^{1/2} = \frac{\mathbf{h}^T \dfrac{d\mathbf{h}}{d\mu}}{\|\mathbf{h}\|}. \qquad (4.11.2)$$

Differentiating equation (4.11.1) gives

$$(M + \mu I)\frac{d\mathbf{h}}{d\mu} = -\mathbf{h}, \qquad (4.11.3)$$

so that

$$\frac{d\mathbf{h}}{d\mu} = (M + \mu I)^{-2}\mathbf{g}. \qquad (4.11.4)$$

This gives

$$\frac{d\|\mathbf{h}\|}{d\mu} = -\frac{\mathbf{g}^T(M + \mu I)^{-3}\mathbf{g}}{\|\mathbf{h}\|}, \qquad (4.11.5)$$

which demonstrates the result as $M + \mu I$ is positively definite for $\mu \geqslant 0$.

(II) The angle ψ which $\mathbf{h}$ makes with $-\mathbf{g}$ decreases monotonically as μ increases.

PROOF.   We have

$$\cos \psi = - \frac{\mathbf{g}^T \mathbf{h}}{\|\mathbf{g}\| \, \|\mathbf{h}\|}, \tag{4.11.6}$$

so that our result is proved if we can show $(d/d\mu)(\cos \psi) > 0$. We have (using equations (4.11.1) to (4.11.9))

$$\frac{d}{d\mu}(\cos \psi) = \frac{-\mathbf{g}^T \dfrac{d\mathbf{h}}{d\mu}}{\|\mathbf{g}\| \, \|\mathbf{h}\|} + \frac{\mathbf{g}^T \mathbf{h}}{\|\mathbf{g}\| \, \|\mathbf{h}\|} \frac{\dfrac{d\|\mathbf{h}\|}{d\mu}}{\|\mathbf{h}\|}$$

$$= \frac{1}{\|\mathbf{g}\| \, \|\mathbf{h}\|^3} \{ -(\mathbf{g}^T(M + \mu I)^{-2}\mathbf{g})^2$$

$$+ (\mathbf{g}^T(M + \mu I)^{-1}\mathbf{g})(\mathbf{g}^T(M + \mu I)^{-3}\mathbf{g}) \}. \tag{4.11.7}$$

To show that the right-hand side of equation (4.11.7) is positive it is sufficient to consider the terms in braces, and to reduce these it is convenient to introduce the spectral decomposition of M which we write

$$M = S^T D S, \tag{4.11.8}$$

where D is diagonal with positive elements and S is orthogonal. As M is positive-definite the elements of D are positive, and we write $\mathbf{v} = S\mathbf{g}$. With this notation the terms in braces in equation (4.11.7) become

$$\sum_{j=1}^{n} \sum_{k=1}^{n} \left\{ - \frac{v_j^2 v_k^2}{(D_j + \mu)^2 (D_k + \mu)^2} + \frac{v_j^2 v_k^2}{(D_j + \mu)(D_k + \mu)^3} \right\}$$

$$= \sum_{j=1}^{n} \sum_{k>j} \frac{v_j^2 v_k^2}{(D_j + \mu)(D_m + \mu)} \left\{ \frac{-2}{(D_j + \mu)(D_k + \mu)} + \frac{1}{(D_j + \mu)^2} + \frac{1}{(D_k + \mu)^2} \right\}$$

$$= \sum_{j=1}^{n} \sum_{k>j} \frac{v_j^2 v_k^2}{(D_j + \mu)(D_k + \mu)} \left\{ \frac{1}{D_j + \mu} - \frac{1}{D_k + \mu} \right\}^2 > 0.$$

(III) The quadratic form $(A\mathbf{x} - \mathbf{f})^T(A\mathbf{x} - \mathbf{f})$ attains its minimum value on the sphere $\|\mathbf{x}\| = \|\mathbf{h}(\mu)\|$ when $\mathbf{x} = \mathbf{h}(\mu)$.

PROOF.   It is readily seen that equation (4.11.1) results if Lagrange multiplier methods are used to solve this problem. Thus $\mathbf{x} = \mathbf{h}(\mu)$ is a stationary value and it remains to show that it is a minimum. This follows

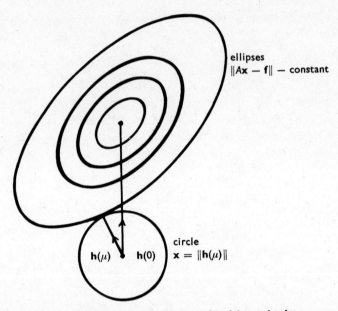

ellipses
$\|Ax - f\|$ — constant

circle
$x = \|h(\mu)\|$

$h(\mu)$ $h(0)$

**Figure 4.11.1.   The minimization problem solved in a single
step of the Marquardt algorithm.**

from Figure 4.11.1. Because $\|h(\mu)\| < \|h(0)\|$ by (I) the situation illustrated
is the only one possible.

REMARK.   (i) The results (I) and (II) show how the algorithm provides
control over the magnitude of h. The result (II) shows that $h(\mu)$ is rotated
toward $-g$ as μ increases. Intuitively this is not necessarily a satisfactory
result (see Figure 4.11.2).

(ii) The comments on the modified Gauss method made in Section 10
apply with equal force here. The vector given by equation (4.11.1) should
be used as a direction of search for the minimum of F.

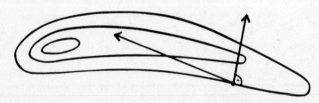

**Figure 4.11.2.   It is not necessarily a disadvantage that the search
direction make a large angle with the negative gradient.**

(iii) Because $\mathbf{h}$ becomes parallel to $\mathbf{g}$ as $\mu \to \infty$ there is no point in taking μ too large, as the resulting method then approximates to the method of steepest descent which is known to be unsatisfactory.

The main problem with this method is the difficulty of experimenting with different values of μ in any step, and it is this that makes the choice of μ difficult. A recent implementation of the algorithm [14] in part avoids this difficulty by making use of the factorization of A described in Section 5. This is based on the observation that equation (4.11.1) is the normal equation for the over-determined system

$$\begin{bmatrix} A \\ \hline \sqrt{\mu}I \end{bmatrix} \mathbf{h} = -\begin{bmatrix} \mathbf{f} \\ 0 \end{bmatrix} + \mathbf{r}. \tag{4.11.9}$$

Using the transformation (4.5.1) this equation can be written

$$\begin{bmatrix} U \\ \hline 0 \\ \hline \sqrt{\mu}I \end{bmatrix} \mathbf{h} = \begin{bmatrix} -W^T\mathbf{f} \\ 0 \end{bmatrix} + \mathbf{s}. \tag{4.11.10}$$

If the methods of Section 4 are now applied to this latter equation we find that the calculation can be carried out conveniently for each of a series of different values of μ because the band of zeros in the matrix of equation (4.11.9) is preserved at each stage of the orthogonal factorization. This is a particular advantage when there are many more equations than there are unknowns, and our analysis of Section 6 has shown that the orthogonal factorization is anyway a perferred method for this type of problem.

It is possible to analyze the manner of convergence of this iteration in the same way as we analyzed the Gauss iteration. Here the correction $\mathbf{h}^{(i)}$ is obtained by solving equation (4.11.8) in the sense of least squares. Whence

$$\begin{bmatrix} \mathbf{f}^{(i)} \\ 0 \end{bmatrix} + \begin{bmatrix} A^{(i)} \\ \hline \sqrt{\mu}I \end{bmatrix} \mathbf{h}^{(i)} = \mathbf{r}^{(i)}, \tag{4.11.11}$$

where

$$[A^{(i)T} \mid \sqrt{\mu}I]\mathbf{r}^{(i)} = 0. \tag{4.11.12}$$

Thus $\mathbf{h}^{(i+1)}$ satisfies, using equations (4.10.3) and (4.11.11),

$$\mathbf{r}^{(i)} + \begin{bmatrix} L^{(i)} \\ \hline -\sqrt{\mu}\mathbf{h}^{(i)} \end{bmatrix} + \begin{bmatrix} A^{(i+1)} \\ \hline \sqrt{\mu}I \end{bmatrix} \mathbf{h}^{(i+1)} = \mathbf{r}^{(i+1)}, \tag{4.11.13}$$

and the corresponding normal equations are

$$[M^{(i+1)} + \mu I]\mathbf{h}^{(i+1)} + [A^{(i+1)T} - A^{(i)T} \mid 0]\mathbf{r}^{(i)}$$

$$+ A^{(i+1)T}\mathbf{L}^{(i)} - \mu\mathbf{h}^{(i)} = 0. \quad (4.11.14)$$

As before, this yields a sufficient condition for the divergence of the unmodified iteration for starting points arbitrarily close to the solution. Here this is

$$\left\| (M + \mu I)^{-1}\left\{\frac{dA^T}{ds}\mathbf{r} - \mu\mathbf{s}\right\} \right\| > 1, \quad (4.11.15)$$

for all directions $\mathbf{s}$ through the solution. It is assumed that all quantities are evaluated at the solution. Here the vector $\mathbf{r}$ is identical with that of equation (4.10.6). It relates only to the first n components of the vector denoted by $\mathbf{r}^{(i)}$ in the preceding equations.

4.12. A METHOD THAT DOES NOT USE DERIVATIVES

It is possible to develop methods for minimizing a general sum of squares without calculating derivatives by a procedure similar to that described in Section 9. Again we keep a matrix of directions, and in terms of this equation (4.9.2) is written

$$\mathbf{f}^{(i)} + A^{(i)T}T^{(i)-1}\mathbf{h}^{(i)} = \mathbf{f}^{(i)} + G^{(i)}\mathbf{a}^{(i)} = \mathbf{r}^{(i)}, \quad (4.12.1)$$

where

$$G_{jk}^{(i)} = \frac{\partial f_j}{\partial t_k}(\mathbf{x}^{(i)}) \quad \text{and} \quad \mathbf{t}_k = \kappa_k(T^{(i)}).$$

As in the secant algorithm the current approximation to G can be "improved" by replacing one of its columns by the result of differencing function values computed during the search for the minimum in the direction generated at the current step of the iteration. In fact we could proceed in exactly similar fashion, and the only major difference would occur in the formulas for updating the inverse of the matrix of the normal equations associated with the current approximation to $G^{(i)}$. In this case this can be done in the following way. We write

$$M^{(i+1)} = G^{(i+1)T}G^{(i+1)}, \quad (4.12.2)$$

where $G^{(i+1)} = G^{(i)} + \mathbf{v}^{(i)}\mathbf{e}_s^T$, and s is the index of the column to be changed. Thus

$$M^{(i+1)} = M^{(i)} + G^{(i)T}\mathbf{v}^{(i)}\mathbf{e}_s^T + \mathbf{v}^{(i)}\rho_s(G^{(i)}) + \mathbf{v}^{(i)T}\mathbf{v}^{(i)}\mathbf{e}_s\mathbf{e}_s^T. \quad (4.12.3)$$

Writing $\tau^{(i)} = \mathbf{v}^{(i)T}\mathbf{v}^{(i)}$ we set

$$Q^{(i)} = M^{(i)} + (G^{(i)T}\mathbf{v}^{(i)} + \tau^{(i)}\mathbf{e}_s)\mathbf{e}_s^T$$

$$= M^{(i)} + \mathbf{q}^{(i)}\mathbf{e}_s^T$$

$$= M^{(i)}\{I + M^{(i)-1}\mathbf{q}^{(i)}\mathbf{e}_s^T\}, \qquad (4.12.4)$$

whence

$$M^{(i+1)} = Q^{(i)}\{I + Q^{(i)-1}\mathbf{v}^{(i)}\rho_s(G^{(i)})\}. \qquad (4.12.5)$$

Thus the inverse of $M^{(i+1)}$ can be found from that of $M^{(i)}$ by multiplying by the inverses of the elementary matrices which appear in equations (4.12.4) and (4.12.5).

The only algorithm of this kind known to us is due to Powell [15]. This differs from the obvious extension of the secant algorithm in certain respects as follows:

(i) The column to be added to $G^{(i)}$ is computed in the form

$$\mathbf{d}^{(i)} = (\mathbf{f}(\mathbf{x}^{(i)} + w\mathbf{t}) - \mathbf{f}(\mathbf{x}^{(i)} + pw\mathbf{t}))/(w - pw) - \mu\mathbf{f}(\mathbf{x}^{(i)} + w\mathbf{t}), \qquad (4.12.6)$$

where $\mathbf{t}$ is the direction of search at the i-th step, w the displacement to the minimum, and pw the displacement to the previously computed point in the one-dimensional search, and where μ is chosen so that

$$\sum_{j=1}^{n} d_j^{(i)} f_j(\mathbf{x}^{(i)} + w\mathbf{t}) = 0. \qquad (4.12.7)$$

This condition is derived by noting that

$$\frac{dF}{d\lambda}(\mathbf{x}^{(i)} + \lambda\mathbf{t}) = 0$$

at the minimum of F in the direction $\mathbf{t}$.

(ii) By equation (4.3.3) the components of the right-hand side of the normal equations associated with the matrix $G^{(i+1)}$ are given by

$$y_j^{(i+1)} = \kappa_j(G^{(i+1)})^T\mathbf{f}^{(i+1)}, \qquad j = 1, 2, \ldots, p. \qquad (4.12.8)$$

By equations (4.12.7) $y_s^{(i+1)} = 0$ where s is the index of the column of $G^{(i)}$ replaced by $\mathbf{d}^{(i)}$. Powell suggests that s be chosen as the index of the maximum of $|a_j^{(i)}y_j^{(i)}|$, $j = 1, 2, \ldots, p$, and this choice ensures that the same index s cannot occur in two consecutive iterations. This choice can be further motivated by noting that $\mathbf{h}^{(i)}$ can become small only if the right-hand side of the normal equations becomes small. This suggests that s be chosen as the index of the component of largest modules of $\mathbf{y}^{(i)}$. However, if the corresponding component of $\mathbf{a}^{(i)}$ is small then we would be replacing a direction that effectively had not been used in the current step. This would tend to make the matrix of directions more nearly singular. Powell's choice can thus be thought of as a compromise.

(iii) Powell scales the matrix of the normal equations so that the diagonal elements are 1. This is achieved by defining $\kappa_s(G^{(i+1)})$ by

$$\kappa_s(G^{(i+1)}) = \mathbf{d}^{(i)}/\|\mathbf{d}^{(i)}\|. \qquad (4.12.9)$$

We have considered an alternative implementation of Powell's algorithm in which the orthogonal factorization technique is used to solve the normal equations. This factorization does not permit the low-cost updating that we have described for $G^{(i)-1}$. However certain economies are possible by noting that some use can be made of the factors obtained in the previous step, for if we write

$$G^{(i)} = \prod_{j=1}^{p} W_j \left[\frac{U}{0} \right]$$

then

$$G^{(i+1)} = \prod_{j=1}^{p} W_j \left[\frac{U}{0} \right] + \mathbf{v}^{(i)} \mathbf{e}_s^T$$

$$= \prod_{j=1}^{s-1} W_j \left\{ \prod_{j=s}^{p} W_j \left[\frac{U}{0} \right] + \prod_{j=p}^{s} W_j \mathbf{v}^{(i)} \mathbf{e}_s^T \right\}, \qquad (4.12.10)$$

and this equation shows that the factorization need only be carried out from column s onward.

REFERENCES

1. R. E. D. Bishop, G. M. L. Gladwell, and S. Michaelson, *The Matrix Analysis of Vibration*, Cambridge University Press, London and New York, 1965, p. 265.
2. J. H. Wilkinson, Error Analysis of Direct Methods of Matrix Inversion, *J. Assoc. Comput. Mach.*, 8(1961), 281.
3. J. H. Wilkinson, *The Algebraic Eigenvalue Problem*, Oxford University Press, London and New York, 1965, p. 191.
4. G. H. Golub, Numerical Methods for Solving Linear Least Squares Problems, *Numer. Math.*, 7(1965), 206
5. G. H. Golub and J. H. Wilkinson, Note on the Iterative Refinement of Least Squares Solution, *Numer. Math.*, 9(1966), 139.
6. M. A. Efroymson, Multiple Regression Analysis, in *Mathematical Methods for Digital Computers*, Vol. 1., A. Ralston and H. S. Wilf, eds., John Wiley & Sons, New York, 1960, p. 191.

7. J. F. Traub. *Iterative Methods for the Solution of Equations*, Prentice-Hall, Englewood Cliffs, N.J., 1964.

8. R. Watts, Research Work in Progress (Solution of a Variant of the Boltzmann Equation), Diffusion Research Unit, Australian National University.

9. J. G. P. Barnes, An Algorithm for Solving Nonlinear Equations Based on the Secant Method, *Computer J.*, *8*(1965), 66.

10. J. Kowalik and J. F. Morrison, Analysis of Kinetic Data for Allosteric Enzyme Reactions as a Nonlinear Regression Problem, *Math. Biosciences*, *2*(1968) 57–66.

11. D. W. Marquardt, An Algorithm for Least Squares Estimation of Nonlinear Parameters, *J. Soc. Indust. Appl. Math.*, *11*(1963), 431.

12. D. D. Morrison, Methods for Nonlinear Least Squares Problems and Convergence Proofs, *JPL Seminar Proceedings*, Space Techn. Lab., Inc., Los Angeles, Calif., 1960.

13. K. Levenberg, A Method for the Solution of Certain Nonlinear Problems in Least Squares, *Quart. Appl. Math.*, *2*(1944), 164.

14. M. Jenkins, research work in progress at computer Science Department, Stanford University, 1967.

15. M. J. D. Powell, A Method for Minimizing a Sum of Squares of Nonlinear Functions without Calculating derivatives, *Computer J.*, *7*(1965), 303.

Chapter 5

CONSTRAINED PROBLEMS

5.1. INTRODUCTION

An important class of applications for the techniques developed in the previous three chapters is to the solution of constrained optimization problems that have been suitably transformed. There are two main types of transformation that can be used for this purpose. First, by suitably transforming the independent variables it may be possible to introduce new variables which are unconstrained. Second, it is possible to transform the objective function by adding severe penalties to it whenever a constraint is violated in such a way that the unconstrained optimization techniques are forced to find minima in the feasible region. In this case the solution is found as the limit of a sequence of solutions to suitably transformed problems. The bulk of this chapter is devoted to the consideration of methods of this second type. These have recently been the subject of intensive study, and some very encouraging results have been obtained in practical applications.

5.2. TRANSFORMATION OF THE INDEPENDENT VARIABLES

Simple transformations, whose application is limited to certain forms of inequality constraints, have been summarized by Box [1]. These transform the independent variables and leaves the objective function unaltered. A simple but frequently encountered practical case of restrictions is that when some or all independent variables are subject to constant lower and upper constraints:

$$l_i \leqslant x_i \leqslant u_i. \tag{5.2.1}$$

If, for example, x_i has to be positive i.e., $l_i = 0, u_i = +\infty$, then the following transformations can be used:

$$x_i = \text{abs}(\bar{x}_i), \qquad x_i = \bar{x}_i^2, \qquad x_i = e^{\bar{x}_i}. \tag{5.2.2}$$

By the transformations

$$x_i = \sin^2 \bar{x}_i \qquad \text{or} \qquad x_i = \frac{e^{\bar{x}_i}}{e^{\bar{x}_i} + e^{-\bar{x}_i}}, \tag{5.2.3}$$

the variable x_i is restricted to the interval defined by $l_i = 0$, $u_i = 1$. In a general case of constant lower and upper constraints (5.2.1) we can apply the transformation

$$x_i = l_i + (u_i - l_i) \sin^2 \bar{x}_i. \tag{5.2.4}$$

After applying any of these transformations the unconstrained optimum of an objective function with respect to the $\bar{x}_i$ variables is sought. It is important to note that such transformations cannot introduce additional and essentially distinct local optima. This is due to the fact that if any point $\bar{x}_i$ is transformed into x_i, then the neighborhood of $\bar{x}_i$ in $\bar{x}$-space is transformed into the neighborhood of x_i in x-space.

5.3. TRANSFORMATION BY THE UNIT STEP FUNCTION

A transformation which utilizes the Heaviside unit step function has been suggested by Courant [2]. This idea has been further developed and successfully implemented in optimum design problems by Schmit and Fox [3]. In this transformation there is a severe penalty imposed on the function each time the feasible region is left, so that the optimizing procedure is forced to backtrack until the point returns to the region where the solution is sought.
We introduce the Heaviside function, which is defined as follows:

$$H(z) = \begin{cases} 1 & \text{for } z < 0, \\ 0 & \text{for } z \geqslant 0, \end{cases} \tag{5.3.1}$$

and consider the constrained optimization problem:

$$\text{minimize } f(\mathbf{x}), \tag{5.3.2}$$

subject to

$$g_i(\mathbf{x}) \geqslant 0, \qquad i = 1, 2, 3, \ldots, I,$$

and

$$e_j(\mathbf{x}) = 0, \qquad j = 1, 1, 3, \ldots, J.$$

In this technique problem (5.3.2) is solved through a sequence of unconstrained minimization problems of the form:

(i) Starting from an arbitrary $\mathbf{x}^{(0)}$, minimize

$$F_1(\mathbf{x}) = (f(\mathbf{x}) - f_1)^2 H(f_1 - f(\mathbf{x})) + \sum_i g_i^2(\mathbf{x})H(g_i(\mathbf{x})) + \sum_j e_j^2(\mathbf{x}), \tag{5.3.3}$$

where f_1 is a preset value such that

$$f_1 < f(\mathbf{x}^{(0)}). \tag{5.3.4}$$

(ii) If the solution $\mathbf{x}^{(1)}$ of (5.3.3) gives

$$\min F_1(\mathbf{x}^{(1)}) = 0, \qquad (5.3.5)$$

then we apply the minimization process to $F_2(\mathbf{x})$, where $f_2 < f_1$, starting from $\mathbf{x}^{(1)}$. Note that any point giving $F_k(\mathbf{x}) = 0$ satisfies the following relationships:

$$f(\mathbf{x}) \leqslant f_k, \qquad (5.3.6)$$

$$g_i(\mathbf{x}) \geqslant 0, \qquad i = 1, 2, \ldots, I,$$

$$e_j(\mathbf{x}) = 0, \qquad j = 1, 2, \ldots, J.$$

(iii) We continue to minimize $F_k(\mathbf{x})$ for a monotonically decreasing sequence of f_k, $k = 1, 2, 3, \ldots$, until for a certain $f_{\bar{k}}$ we obtain

$$\min F_{\bar{k}}(\mathbf{x}) > 0, \qquad (5.3.7)$$

which would indicate that

$$f_{\bar{k}-1} \geqslant \min f(\mathbf{x}) > f_{\bar{k}}. \qquad (5.3.8)$$

A more accurate solution than $\mathbf{x}^{(\bar{k}-1)}$ can be found by repeating the minimization procedure for f_k values, chosen in the interval $(f_{\bar{k}-1}, f_{\bar{k}})$.

In practical applications of transformation (5.3.3) we introduce scaling factors λ_i to ensure that the contributions of the constraints are of the same order of magnitude. Thus the transformation is

$$F_k(\mathbf{x}) = (f(\mathbf{x}) - f_k)^2 H(f_k - f(\mathbf{x})) + \sum_i \lambda_i g_i^2(\mathbf{x}) H(g_i(\mathbf{x})) + \sum_j \lambda_j e_j^2(\mathbf{x}). \qquad (5.3.9)$$

It is also necessary to introduce a less restrictive criterion than $\min F_k(\mathbf{x}) = 0$ to terminate the minimization in the k-th step. If we define

$$G_k(\mathbf{x}) = \max \{(f(\mathbf{x}) - f_k)^2 H(f_k - f(\mathbf{x})), \lambda_i g_i^2(\mathbf{x}) H(g_i(\mathbf{x})), \lambda_j e_j^2(\mathbf{x})\}, \qquad (5.3.10)$$

then a possible criterion is

$$G_k(\mathbf{x}) \leqslant \delta, \qquad (5.3.11)$$

where δ is a preset positive number.

5.4. CARROLL'S CREATED RESPONSE SURFACE TECHNIQUE

Carroll [4] has suggested that the solution to the constrained minimization problem,

$$\text{minimize } f(\mathbf{x}), \qquad (5.4.1)$$

subject to the constraints

$$g_i(\mathbf{x}) \geqslant 0, \qquad i = 1, 2, \ldots I,$$

might be found as the limit as $r_k \to 0$ of the solutions to the unconstrained minimization problem

$$\text{minimize } T(\mathbf{x}, r_k) = f(\mathbf{x}) + r_k F(g_i(\mathbf{x}), \text{ all } i), \tag{5.4.2}$$

where the function $F(\mathbf{g})$ has the following properties:

(i) $F(\mathbf{g})$ is defined and bounded below for all $g_i(\mathbf{x}) > 0$, and
(ii) $F(\mathbf{g}) \to \infty$ if any $g_i(\mathbf{x}) \to 0$.

The perturbation parameter $r_k > 0$ is monotonically decreasing, with

$$\lim r_k = 0, \qquad k \to \infty. \tag{5.4.3}$$

Carroll chooses the simplest form of $F(\mathbf{g})$:

$$F(\mathbf{g}) = \sum_i g_i^{-1}(\mathbf{x}). \tag{5.4.4}$$

However, other alternatives are possible, e.g.,

$$F(\mathbf{g}) = -\sum_i \log (g_i(\mathbf{x})). \tag{5.4.5}$$

For further considerations we introduce the following definitions:

1. A point is called feasible if it belongs to a feasible set (region)

$$R = \{\mathbf{x} \mid g_i(\mathbf{x}) \geqslant 0, \, i = 1, 2, \ldots, I\}. \tag{5.4.6}$$

Any feasible point satisfies the constraints of problem (5.4.1).
2. A point is called an interior-feasible point if it belongs to a set

$$R^0 = \{\mathbf{x} \mid g_i(\mathbf{x}) > 0, \, i = 1, 2, \ldots, I\}. \tag{5.4.7}$$

The sets R and R^0 are called feasible and interior-feasible, respectively.
3. The point $\mathbf{x}_B$ is called a boundary point if every neighborhood of $\mathbf{x}_B$ defined by $\|\mathbf{x} - \mathbf{x}_B\| < \varepsilon$, where $\varepsilon > 0$, contains points belonging to R^0 and also points violating at least one constraint.

Let us assume that we know at least one point $\mathbf{x}^{(0)} \in R^0$ and that we can use one of the efficient unconstrained optimization techniques. The following procedure is then possible:

Starting from $\mathbf{x}^{(0)}$ decrease the value of $T(\mathbf{x}, r_1)$ where

$$T(\mathbf{x}, r_1) = f(\mathbf{x}) + r_1 \sum_i g_i^{-1}(\mathbf{x}) \tag{5.4.8}$$

and $r_1 > 0$. Suppose that the feasible region R is never left. For convenience we can assume that our minimization method defines a continuous n-dimensional curve on which $T(\mathbf{x}, r_1)$ is decreasing. Moving along this curve we should be able to find the point $\mathbf{x}(r_1) \in R^0$, which minimizes $T(\mathbf{x}, r_1)$. Clearly the boundary can never be violated because $T(\mathbf{x}, r_1) \to \infty$, as the boundary is approached so that the minimum $\mathbf{x}(r_1)$ (if it exists) must belong

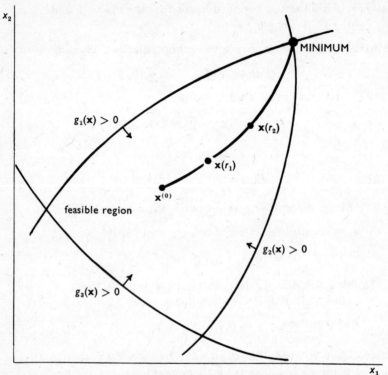

Figure 5.4.1.   Sequence of solutions of a hypothetical two-dimensional problem.

to the interior-feasible region. This procedure is repeated for $0 < r_{k+1} < r_k$, $k = 1, 2, \ldots$ and the minimization of $T(\mathbf{x}, r_k)$ can be achieved without violating the constraints. Ultimately it is hoped that a sequence of the $f(\mathbf{x}(r_k))$ will converge to the minimum of $f(\mathbf{x})$ as the value of r_k tends to zero. Due to the structure of $T(\mathbf{x}, r_k)$ all the solutions $\mathbf{x}(r_k) \in R^0$ and none can satisfy strictly any of inequalities $g_i(\mathbf{x})$. However, we can expect that by taking r_k small enough we should be able to approach as close to the boundary as we wish.

To illustrate this procedure, consider a simple two-dimensional problem (Figure 5.4.1). We have to determine $\mathbf{x}$ which minimizes the objective function $f(\mathbf{x})$ and satisfies the three constraints $g_i(\mathbf{x}) \geqslant 0$, $i = 1, 2, 3$. Starting from $\mathbf{x}^{(0)} \in R^0$ and setting $r = r_1 > 0$ the minimization produces the point $\mathbf{x}(r_1)$ which is the minimum of $T(\mathbf{x}, r_1)$. We then repeat the minimization by selecting r_2, where $0 < r_2 < r_1$, and starting from $\mathbf{x}(r_1)$. By continuing the process a sequence of minima is obtained, and we expect that under suitable conditions

$$\lim_{r_k \to 0} f(\mathbf{x}(r_k)) = \min_{\mathbf{x} \in R} f(\mathbf{x}). \qquad (5.4.9)$$

Figure 5.4.1 also shows that not all constraints are equally important. The constraints $g_1(\mathbf{x})$, $g_2(\mathbf{x})$ are active and more likely to be violated than the constraint $g_3(\mathbf{x})$. This would suggest that we can modify our computational procedure starting the minimization with the untransformed objective function, and when any constraint is violated we return to the last interior-feasible point and transform the objective function by allowing for the contribution from the violated constraint. It should also be possible to drop constraints when the current point moves away from them.

To illustrate numerical aspects of the Carroll transformation, consider the simple nonlinear problem,

$$\text{minimize } f(\mathbf{x}) = x_1 + x_2, \qquad (5.4.10)$$

subject to the constraint

$$4x_1 + 4x_2 - x_1^2 - x_2^2 - 7 \geqslant 0.$$

The reasible region for this problem is the interior of a disk of unit radius centered at $(2, 2)$ (see Figure 5.4.2). The solution is $\mathbf{x} = (2 - 1/\sqrt{2}, 2 - 1/\sqrt{2})$ and $f(\mathbf{x}) = 4 - \sqrt{2}$. The transformed function for this problem is:

$$T(\mathbf{x}, r) = x_1 + x_2 + r(4x_1 + 4x_2 - x_1^2 - x_2^2 - 7)^{-1}. \qquad (5.4.11)$$

The necessary conditions for an optimum of $T(\mathbf{x}, r)$ are;

$$\frac{\partial T}{\partial x_1} = 1 - r(4 - 2x_1)(4x_1 + 4x_2 - x_1^2 - x_2^2 - 7)^{-2} = 0, \qquad (5.4.12)$$

$$\frac{\partial T}{\partial x_2} = 1 - r(4 - 2x_2)(4x_1 + 4x_2 - x_1^2 - x_2^2 - 7)^{-2} = 0. \qquad (5.4.13)$$

From these equations we see that $x_1(r) = x_2(r)$. Solving one of them we have

$$x_1(r) = 2 + r/2 \pm 0.25\sqrt{4r^2 + 8}, \qquad (5.4.14)$$

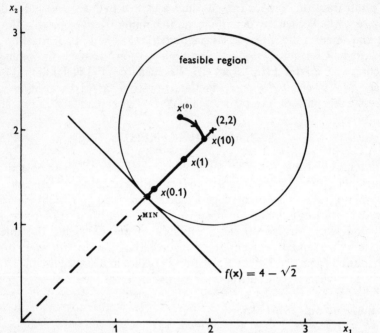

Figure 5.4.2. Minimum of T(x, r) of the nonlinear problem (5.4.10).

where the negative second term corresponds to the minimum. The solution of (5.4.10) should be approached by the sequence of $\mathbf{x}(r)$ as $r \rightarrow 0$. We have therefore

$$x_1^{\mathrm{MIN}} = x_2^{\mathrm{MIN}} = \lim_{r \rightarrow 0} (2 + r/2 - 0.25\sqrt{4r^2 + 8})$$

$$= 2 - 1/\sqrt{2} = 1.29289, \qquad (5.4.15)$$

and

$$f^{\mathrm{MIN}} = f(\mathbf{x}^{\mathrm{MIN}}) = 4 - \sqrt{2}, \qquad (5.4.16)$$

which is in agreement with the correct solution.

Let us suppose that we have started the process of minimization with $r = 10$ from $\mathbf{x}^{(0)}$ (Figure 5.4.2). Using a suitable minimization technique we should be able to find the solution $\mathbf{x}(10) = (1.95024, 1.95024)$. By continuing the process the trajectory of minima is generated which is the straight line $x_1 = x_2$. Table 5.4.1 gives the subsequent values of $\mathbf{x}(r)$ for a decreasing sequence of values of r. We see that for r small enough (say $r = 0.0001$) the

Table 5.4.1

r	$x_1(r) = x_2(r)$
10	1.95024
1	1.63397
.1	1.34112
.0001	1.29294
0	1.29289

method produces quite a good approximation to the correct solution $x(0) = (1.29289, 1.29289)$.

5.5. CONVEX PROBLEM I

In this and the next paragraphs we shall present the theoretical validation and development of Carroll's method which are due to Fiacco and McCormick [5]. Two distinct problems are considered: Problem I, with inequality constraints only, and Problem II, which includes also strict equality constraints.

In order to discuss the theoretical aspects of the transformed objective function $T(\mathbf{x}, r_k)$ we have to restrict the class of problems under consideration. The conditions we now impose on the optimization problem are not severe and are often satisfied in practice. These are:

(A) The interior-feasible set R^0 is not empty and we can find one point $\mathbf{x}^{(0)} \in R^0$.

(B) The objective and constraining functions belong to the class C^2 (twice continuously differentiable).

(C) The objective function $f(\mathbf{x})$ is bounded below on R, i.e., we assume the existence of a greatest lower bound L_f such that

$$f(\mathbf{x}) \geqslant L_f > -\infty. \tag{5.5.1}$$

(D) The set $S = \{\mathbf{x} \,|\, f(\mathbf{x}) \leqslant K, \mathbf{x} \in R\}$ is bounded for every finite value of K.

Conditions (A) to (C) are self-explanatory, while condition (D) ensures that any local minimum is located at a finite point.

It is clear that under the conditions (A) to (D) the problem

$$\underset{x \in R^0}{\text{minimize }} T(\mathbf{x}, r_k) = f(\mathbf{x}) + r_k \sum_i g_i^{-1}(\mathbf{x}), \tag{5.5.2}$$

has at least one local minimum $\mathbf{x}(r_k)$ which is finite. For this point $\nabla T(\mathbf{x}(r_k), r_k)$ vanishes and the Hessian is positive semi-definite.

To ensure the existence of a global minimum of $T(\mathbf{x}, r_k)$ we have to introduce some additional conditions. It is sufficient that $T(\mathbf{x}, r_k)$ be a strictly convex function of $\mathbf{x}$ over the set R^0 for every positive value of the perturbation parameter (see Appendix 2). Since

$$T(\mathbf{x}, r_k) = f(\mathbf{x}) + r_k \sum_i g_i^{-1}(\mathbf{x}), \qquad r_k > 0, \tag{5.5.3}$$

it is sufficient to assume that:

(E) $f(\mathbf{x})$ and $g_i^{-1}(\mathbf{x})$ are convex (which implies that the $g_i(\mathbf{x})$ are concave).
(F) At least one of $f(\mathbf{x})$, $g_i^{-1}(\mathbf{x})$ is strictly convex.

These conditions of convexity are essential to our argument but are difficult to verify and certainly do not hold in general. However, when conditions (A) to (F) are fulfilled then our problem is to minimize the strictly convex function $T(\mathbf{x}, r_k)$, where $\mathbf{x} \in R^0$ which is a convex set. It follows that for every $r_k > 0$ there exists a unique minimum of $T(\mathbf{x}, r_k)$. We show first that if $0 < r_s < r_g$ then

$$T(\mathbf{x}(r_s), r_s) < T(\mathbf{x}(r_g), r_g), \tag{5.5.4}$$

where $\mathbf{x}(r_s)$ minimizes $T(\mathbf{x}, r_s)$ and $\mathbf{x}(r_g)$ minimizes $T(\mathbf{x}, r_g)$. We have

$$T(\mathbf{x}(r_g), r_g) = f(\mathbf{x}(r_g)) + r_g \sum_i g_i^{-1}(\mathbf{x}(r_g))$$

$$> f(\mathbf{x}(r_g)) + r_s \sum_i g_i^{-1}(\mathbf{x}(r_g))$$

$$> T(\mathbf{x}(r_s), r_s). \tag{5.5.5}$$

We can now prove the following Convergence Theorem for convex Problem I.

THEOREM 5.5.1. *Under the assumed conditions* (A) *to* (F),

$$\lim_{r_k \to 0} \min_{\mathbf{x}} T(\mathbf{x}, r_k) = \min_{\mathbf{x} \in R} f(\mathbf{x}) \overset{\text{def}}{\equiv} L_f, \tag{5.5.6}$$

i.e., the values $T(\mathbf{x}(r_k), r_k)$, $k = 1, 2, 3, \ldots$, *approach the solution of the constrained convex Problem I as the perturbation parameter tends to zero.*

PROOF. To prove the theorem it is sufficient to show that for any $\delta > 0$ there is a $k(\delta)$ such that if $k > k(\delta)$ $(r_k < r_{k(\delta)})$ then

$$|\min_{\mathbf{x}} T(\mathbf{x}, r_k) - L_f| < \delta. \tag{5.5.7}$$

Since $f(\mathbf{x})$ is continuous it is possible to select $\bar{\mathbf{x}} \in R^0$ such that

$$f(\bar{\mathbf{x}}) < L_f + \delta/2. \tag{5.5.8}$$

Choose $r_{k(\delta)}$ such that

$$r_{k(\delta)} \max_i g_i^{-1}(\bar{\mathbf{x}}) < \delta/2I, \tag{5.5.9}$$

where I is the number of constraining inequalities. By relationship (5.5.5) we obtain, if $r_k < r_{k(\delta)}$,

$$T(\mathbf{x}^{(k)}, r_k) < T(\mathbf{x}^{(k(\delta))}, r_{k(\delta)}), \tag{5.5.10}$$

where $\mathbf{x}^{(k)} = \mathbf{x}(r_k)$ and $\mathbf{x}^{(k(\delta))} = \mathbf{x}(r_{k(\delta)})$. Furthermore

$$T(\mathbf{x}^{(k(\delta))}, r_{k(\delta)}) < T(\bar{\mathbf{x}}, r_{k(\delta)}) < L_f + \delta. \tag{5.5.11}$$

On the other hand,

$$T(\mathbf{x}^{(k)}, r_k) > f(\mathbf{x}^{(k)}) \geqslant L_f > L_f - \delta. \tag{5.5.12}$$

Combining the inequalities (5.5.10)–(5.5.12) we have

$$L_f - \delta < T(\mathbf{x}^{(k)}, r_k) < L_f + \delta, \tag{5.5.13}$$

which is equivalent to

$$\left| \min_{\mathbf{x}} T(\mathbf{x}, r_k) - L_f \right| < \delta, \tag{5.5.14}$$

and this proves the theorem.

It follows that when $r_k \to 0$, we have

$$\lim f(\mathbf{x}^{(k)}) = L_f, \tag{5.5.15}$$

and

$$\lim r_k \sum_i g_i^{-1}(\mathbf{x}^{(k)}) = 0. \tag{5.5.16}$$

An equivalence has been shown between the convex problem I and the minimization of $T(\mathbf{x}, r_k)$, for a strictly decreasing sequence of r_k tending to zero. It can also be shown that if r_k, $k = 1, 2, 3, \ldots$, is a strictly decreasing sequence of positive values then the sequence $f(\mathbf{x}^{(k)})$ is also strictly decreasing. Let us assume that $0 < r_{k+1} < r_k$; then we can combine the following two relationships,

$$f(\mathbf{x}^{(k+1)}) + r_{k+1} \sum_i g_i^{-1}(\mathbf{x}^{(k+1)}) < f(\mathbf{x}^{(k)}) + r_{k+1} \sum_i g_i^{-1}(\mathbf{x}^{(k)}), \tag{5.5.17}$$

$$f(\mathbf{x}^{(k)}) + r_k \sum_i g_i^{-1}(\mathbf{x}^{(k)}) < f(\mathbf{x}^{(k+1)}) + r_k \sum_i g_i^{-1}(\mathbf{x}^{(k+1)}), \tag{5.5.18}$$

to obtain

$$f(\mathbf{x}^{(k+1)})(r_{k+1}^{-1} - r_k^{-1}) < f(\mathbf{x}^{(k)})(r_{k+1}^{-1} - r_k^{-1}), \tag{5.5.19}$$

whence

$$f(\mathbf{x}^{(k+1)}) < f(\mathbf{x}^{(k)}). \tag{5.5.20}$$

From equations (5.5.17) and (5.5.18) we can also derive the result

$$\sum_i g_i^{-1}(\mathbf{x}^{(k+1)}) > \sum_i g_i^{-1}(\mathbf{x}^{(k)}). \tag{5.5.21}$$

REMARK.   When the convexity conditions are not satisfied, or when the problem is of a considerable complexity and we do not know beforehand whether these conditions are satisfied, then it is impossible to guarantee that a local solution is also a global one. In general, if the convexity requirements are removed, the SUMT method can be used to find a local minimum. These difficulties may be encountered by any method and although theoretically serious have not caused great troubles in many practical applications. By systematically starting the iterations from different initial points it is frequently possible to locate the global solution even if several optima exist. Of course, it is necessary to be able to recognize the desired optimum, and in practice this means that there is no substitute for understanding the problem.

5.6.   CONVEX PROBLEM II

Carroll's transformation cannot be applied directly to solving problems with strict equality constraints. In order to handle the more general class of problems

$$\text{minimize } f(\mathbf{x}) \tag{5.6.1}$$

subject to

$$g_i(\mathbf{x}) \geqslant 0, \qquad i = 1, 2, \ldots, I,$$

and

$$e_j(\mathbf{x}) = 0, \qquad j = 1, 2, \ldots, J,$$

the following modification of $T(\mathbf{x}, r_k)$ can be used

$$T(\mathbf{x}, r_k) = f(\mathbf{x}) + r_k F(\mathbf{g}) + G(r_k) \|\mathbf{e}\|^2, \tag{5.6.2}$$

where $G(r_k) \to \infty$, when $r_k \to 0$. The motivation for the third term in (5.6.2) is that as $G(r_k) \to \infty$ the norm $\|\mathbf{e}\|^2 = \sum_j e_j^2(\mathbf{x})$ must tend to zero, otherwise $T(\mathbf{x}, r_k)$ would increase to infinity, which cannot occur when a stable minimization procedure is used. The following variant of transformation (5.6.2) has been used by Fiacco and McCormick [6].

$$T(\mathbf{x}, r_k) = f(\mathbf{x}) + r_k \sum_i g_i^{-1}(\mathbf{x}) + r_k^{-1/2} \sum_j e_j^2(\mathbf{x}). \tag{5.6.3}$$

We introduce the new set $R^1 = \{\mathbf{x} \mid e_j(\mathbf{x}) = 0, \text{ all } j\}$; the sets R^0 and R are defined as before. In order to make Problem II well-behaved and convex a number of restrictions need to be imposed. These conditions are analogous to those of Problem I and reduce to them when equality constraints do not appear.

The assumptions additional to those in Section 5 ((A) to (F)) are:

(i) $R^0 \cap R^1$ is nonempty,

(ii) $\sum_j e_j^2(\mathbf{x})$ is convex for $\mathbf{x} \in R^0$.

(iii) $T(\mathbf{x}, r_k)$ defined by (5.6.3) is strictly convex in R^0 for every $r_k > 0$.

Under these conditions the following theorem [7] can be proved by an argument similar to that employed for Theorem 5.5.1. The details are left to the reader.

THEOREM 5.6.1.   For each $r_k > 0$ let $\mathbf{x}^{(k)}$ be the unique minimum of $T(\mathbf{x}, r_k)$, then the values $T(\mathbf{x}^{(k)}, r_k)$, $k = 1, 2, \ldots$, approach the solution of the original convex Problem II (5.6.1) as $r_k \to 0$, and

$$\lim_{r_k \to 0} \min T(\mathbf{x}, r_k) = \min_{\mathbf{x} \in R \cap R^1} f(\mathbf{x}) = L_f. \tag{5.6.4}$$

Obviously we have

$$\nabla T(\mathbf{x}^{(k)}, r_k) = 0, \tag{5.6.5}$$

for each $\mathbf{x}^{(k)}$.

5.7. DUAL PROBLEMS

Consider the SUMT (sequential unconstrained minimization technique) transformation of Problem I and its corresponding dual formulation. We assume the validity of conditions (A)–(F). We know that if $\mathbf{x}^{(k)} = \mathbf{x}(r_k)$ is a point which minimizes $T(\mathbf{x}, r_k)$ then

$$\nabla T(\mathbf{x}^{(k)}, r_k) = \nabla f(\mathbf{x}^{(k)}) - r_k \sum_i g_i^{-2}(\mathbf{x}^{(k)}) \nabla g_i(\mathbf{x}^{(k)}) = 0. \tag{5.7.1}$$

Choosing

$$v_i = r_k g_i^{-2}(\mathbf{x}^{(k)}) > 0, \tag{5.7.2}$$

we obtain from (5.7.1)

$$\nabla f(\mathbf{x}^{(k)}) = \sum_i v_i \nabla g_i(\mathbf{x}^{(k)}), \tag{5.7.3}$$

which means that any pair $(\mathbf{x}^{(k)}, \mathbf{v})$ satisfies the side constraints of the dual problem. It follows from the Convergence Theorem 5.5.1 that $\lim f(\mathbf{x}^{(k)}) = L_f$ and

$$\lim r_k \sum_i g_i^{-1}(\mathbf{x}^{(k)}) = \lim \sum_i v_i g_i(\mathbf{x}^{(k)}) = 0$$

as $r_k \to 0$. Therefore, if δ is any positive value, then $r_{k(\delta)}$ can be found such that for $r_k < r_{k(\delta)}$ we have

$$L_f \leqslant f(\mathbf{x}^{(k)}) < L_f + \delta, \qquad (5.7.4)$$

$$-\delta < -r_k \sum_i g_i^{-1}(\mathbf{x}^{(k)}) < 0, \qquad (5.7.5)$$

and finally we obtain

$$|D(\mathbf{x}^{(k)}, v) - L_f| < \delta, \qquad (5.7.6)$$

where

$$D(\mathbf{x}^{(k)}, \mathbf{v}) = f(\mathbf{x}^{(k)}) - \sum_i v_i g_i(\mathbf{x}^{(k)}), \qquad (5.7.7)$$

and v_i is defined by (5.7.2). These results can be summarized as follows: the SUMT method produces points $(\mathbf{x}^{(k)}, \mathbf{v}(r_k))$, where $v_i = r_k g_i^{-1}(\mathbf{x}^{(k)})$, which are feasible for the dual problem, and values $D(\mathbf{x}^{(k)}, \mathbf{v}(r_k))$, which tend to L_f as r_k tends to zero. L_f is the maximum value of D and consequently we have

$$D(\mathbf{x}^{(k)}, \mathbf{v}) = f(\mathbf{x}^{(k)}) - r_k \sum_i g_i^{-1}(\mathbf{x}^{(k)}) \leqslant L_f < f(\mathbf{x}^{(k)}). \qquad (5.7.8)$$

This inequality has an important practical value because it provides upper and lower bounds on the final solution L_f and gives an easy convergence criterion.

A similar dual formulation may be constructed for the convex Problem II with the equality constraints. However, unless the equality constraints $e_j(x) = 0$ are linear, the solution of the corresponding dual problem can give a value which is not necessarily equal to L_f, the solution to the primal problem. Nevertheless, the method yields points which are dual-feasible and whose corresponding D values bound L_f from below. A full similarity to the solution of the dual formulation of Problem I has been shown [7] only for the case when all $e_j(\mathbf{x})$ are linear.

5.8.   SUMT WITHOUT PERTURBATION PARAMETER

Fiacco and McCormick [8] have suggested an interesting approach to Problem I which does not makes use of the parameters r_k. This variation of the SUMT transformation produces a strictly decreasing sequence of values of the function $f(\mathbf{x})$.

It has been shown that for the T-transformation we have

$$T(\mathbf{x}(r_k), r_k) > T(\mathbf{x}(r_{k+1}), r_{k+1}), \qquad (5.8.1)$$

and

$$f(\mathbf{x}(r_k)) > f(\mathbf{x}(r_{k+1})), \qquad (5.8.2)$$

where $r_k > r_{k+1} > 0$, $k = 1, 2, \ldots$. It is, however, not necessarily true that

$$f(\mathbf{x}^{(0)}) > f(\mathbf{x}(r_1)), \qquad (5.8.3)$$

since the starting point and the value of r_1 are independent and arbitrary. Here the following W-transformation replaces the previously constructed T-transformation:

$$W(\mathbf{x}, \mathbf{x}^{(k)}) = [f(\mathbf{x}^{(k)}) - f(\mathbf{x})]^{-1} + \sum_i g_i^{-1}(\mathbf{x}). \qquad (5.8.4)$$

The computational procedure is:

(i) Start with $\mathbf{x}^{(0)} \in R^0$.

(ii) Suppose that the first k points $\mathbf{x}^{(0)}, \mathbf{x}^{(1)}, \ldots, \mathbf{x}^{(k-1)}$ (all in R^0) have been obtained. Then we minimize $W(\mathbf{x}, \mathbf{x}^{(k-1)})$ in the region $\{\mathbf{x} \mid f(\mathbf{x}) < f(\mathbf{x}^{(k-1)})\}$ and $(\mathbf{x} \in R^0)$.

We start from a point that must be different from $\mathbf{x}^{(k-1)}$ to ensure that $W(\mathbf{x}, \mathbf{x}^{(k-1)})$ is finite. One possibility is

$$\mathbf{x} = \mathbf{x}^{(k-1)} - \lambda \nabla f(\mathbf{x}^{(k-1)}), \qquad (5.8.5)$$

where $\lambda > 0$ is small enough to assure that $\mathbf{x} \in R^0$. Under suitable conditions which are analogous to those required for the T-transformation the sequence of points $\mathbf{x}^{(k)}$ tends to the solution of Problem I.

We introduce $R_k = \{\mathbf{x} \mid f(\mathbf{x}) \leqslant f(\mathbf{x}^{(k)})\}$, $k = 1, 2, \ldots$, and $R_k^0 = \{\mathbf{x} \mid f(\mathbf{x}) < f(\mathbf{x}^{(k)})\}$. Sets R and R^0 are defined as before.

We assume that the following conditions are imposed on the problem:

(A) The set R^0 is nonempty.

(B) The functions $\{f(\mathbf{x}), g_i(\mathbf{x}), \text{all } i\} \in C^2$.

(C) The set $R \cap R_k$ is bounded for every finite k.

(D) The functions $f(\mathbf{x})$, $g_i^{-1}(\mathbf{x})$ are convex.

(E) The transformation function $W(\mathbf{x}, \mathbf{x}^{(k)})$ defined by (5.8.4) is strictly convex in $R^0 \cap R_k^0$ for every k.

Conditions (A) to (C) imply the existence of a finite number L_f such that

$$L_f = \min_{x \in R} f(\mathbf{x}). \qquad (5.8.6)$$

The last condition ensures uniqueness of the minimum of the W-function and is satisfied if $f(\mathbf{x})$ or any $g_i^{-1}(\mathbf{x})$ is strictly convex.

Under the above conditions, the following convergence theorem has been proved:

THEOREM 5.8.1.

(1) *Let* $\mathbf{x}^{(k)}$ *minimize* $W(\mathbf{x}, \mathbf{x}^{(k-1)})$, *and let* $R^0 \cap R_k^0$ *be nonempty, then the sequence of values* $f(\mathbf{x}^{(k)})$, $k = 1, 2, \ldots$, *tends to* L_f *as* $k \to \infty$.

(2) *If for some finite* k *the set* $R^0 \cap R_k^0$ *is empty then*

$$f(\mathbf{x}^{(k)}) = \min_{x \in R} f(\mathbf{x}) = L_f, \tag{5.8.7}$$

and

$$\nabla f(\mathbf{x}^{(k)}) = \mathbf{0}, \tag{5.8.8}$$

i.e., $\mathbf{x}^{(k)}$ *is the unconstrained minimum of* $f(\mathbf{x})$.

Another important theorem gives a simple and explicit relationship between the problems of minimizing the T and W functions:

THEOREM 5.8.2. *Under the assumed conditions* (A) *to* (E) *the point* $\mathbf{x}^{(k+1)}$ *minimizes* $W(\mathbf{x}, \mathbf{x}^{(k)})$ *and also* $T(\mathbf{x}, r_k)$ *if*

$$r_k = d_k^2 = [f(\mathbf{x}^{(k)}) - f(\mathbf{x}^{(k+1)})]^2. \tag{5.8.9}$$

PROOF. A direct computation shows that

$$0 = \nabla W(\mathbf{x}^{(k+1)}, \mathbf{x}^{(k)}) = d_k^{-2} \nabla T(\mathbf{x}^{(k+1)}, d_k^2), \tag{5.8.10}$$

whence $\mathbf{x}^{(k+1)}$ is also the minimum of $T(\mathbf{x}, d_k^2)$.

We shall also show that the sequence of values, d_k, $k = 0, 1, 2, \ldots$, is strictly decreasing, i.e.,

$$d_k = f(\mathbf{x}^{(k)}) - f(\mathbf{x}^{(k+1)}) > f(\mathbf{x}^{(k+1)}) - f(\mathbf{x}^{(k+2)}) = d_{k+1}. \tag{5.8.11}$$

Bearing in mind that $\min T(\mathbf{x}, d_k^2) = T(\mathbf{x}^{(k+1)}, d_k^2)$ we have the following two inequalities

$$f(\mathbf{x}^{(k+1)}) + d_k^2 \sum_i g_i^{-1}(\mathbf{x}^{(k+1)}) < f(\mathbf{x}^{(k+2)}) + d_k^2 \sum_i g_i^{-1}(\mathbf{x}^{(k+2)}), \tag{5.8.12}$$

$$f(\mathbf{x}^{(k+2)}) + d_{k+1}^2 \sum_i g_i^{-1}(\mathbf{x}^{(k+2)}) < f(\mathbf{x}^{(k+1)}) + d_{k+1}^2 \sum_i g_i^{-1}(\mathbf{x}^{(k+1)}). \tag{5.8.13}$$

Adding these inequalities, we obtain

$$(d_k^2 - d_{k+1}^2)\left[\sum_i g_i^{-1}(\mathbf{x}^{(k+2)}) - \sum_i g_i^{-1}(\mathbf{x}^{(k+1)})\right] > 0. \tag{5.8.14}$$

We know also that

$$f(\mathbf{x}^{(k)}) > f(\mathbf{x}^{(k+1)}) > f(\mathbf{x}^{(k+2)}), \tag{5.8.15}$$

which gives

$$f(\mathbf{x}^{(k)}) - f(\mathbf{x}^{(k+2)}) > f(\mathbf{x}^{(k)}) - f(\mathbf{x}^{(k+1)}). \tag{5.8.16}$$

The function $W(\mathbf{x}, \mathbf{x}^{(k)})$ is minimized at $\mathbf{x}^{(k+1)}$ and therefore

$$W(\mathbf{x}^{(k+1)}, \mathbf{x}^{(k)}) < W(\mathbf{x}^{(k+2)}, \mathbf{x}^{(k)}), \qquad (5.8.17)$$

that is,

$$[f(\mathbf{x}^{(k)}) - f(\mathbf{x}^{(k+1)})]^{-1} + \sum_i g_i^{-1}(\mathbf{x}^{(k+1)})$$

$$< [f(\mathbf{x}^{(k)}) - f(\mathbf{x}^{(k+2)})]^{-1} + \sum_i g_i^{-1}(\mathbf{x}^{(k+2)}). \quad (5.8.18)$$

Combining (5.8.16) and (5.8.18) it follows that

$$\sum_i g_i^{-1}(\mathbf{x}^{(k+2)}) - \sum_i g_i^{-1}(\mathbf{x}^{(k+1)}) > 0. \qquad (5.8.19)$$

We see that the second term of inequality (5.8.14) is positive and therefore

$$d_k^2 - d_{k+1}^2 > 0, \qquad (5.8.20)$$

which finally gives

$$d_k > d_{k+1}, k = 0, 1, 2, \ldots . \qquad (5.8.21)$$

This strictly decreasing sequence of d_k corresponds to the property of $T(\mathbf{x}, r_k)$ which is minimized for a strictly decreasing sequence of r_k. It is also obvious that

$$\lim_{k \to \infty} d_k = 0. \qquad (5.8.22)$$

Since $x(r_k)$ converges to the solution when r_k is a strictly decreasing sequence of positive values with $\lim r_k = 0$, it follows that d_k^2 is a particular realization of this sequence of r_k.

REMARK. The main difference between the T and W transformations is that, once the starting point $\mathbf{x}^{(0)}$ is selected, the entire process of minimizing $W(\mathbf{x}, \mathbf{x}^{(k)})$ is fixed, but this is not the case when T-transformation is used. For the W-transformation we have

$$f(\mathbf{x}^{(0)}) > f(\mathbf{x}^{(1)}) > f(\mathbf{x}^{(2)}) > \cdots \text{etc.}, \qquad (5.8.23)$$

while the T-transformation ensures only that

$$f(\mathbf{x}^{(1)}) > f(\mathbf{x}^{(2)}) > \cdots \text{etc.} \qquad (5.8.24)$$

These two properties of the W-transformation may be considered as advantages. However, the fact that the perturbation parameter r_k disappears from $W(\mathbf{x}, \mathbf{x}^{(k)})$ precludes any possibility of controlling the process of minimization.

5.9. SELECTION OF FEASIBLE POINT AND PERTURBATION PARAMETER

In order to compute the SUMT transformation we have to know at least one interior point $x^{(0)}$. In many practical applications such a point is easily constructed from a knowledge of the problem. However, it may happen that an initial point is not available, so that it is necessary to solve an auxiliary problem to find one.

The possibility of using the SUMT method to find an $x^{(0)} \in R^0$ has been suggested by Fiacco [9]. Suppose that a point $x^{(0)}$ is given which does not satisfy all the constraints $g_i(x) > 0$. Introduce two sets of indices, $I_1 = \{i \mid g_i(x^{(0)}) > 0\}$ and $I_2 = \{i \mid g_i(x^{(0)}) \leqslant 0\}$.

This procedure for finding an interior-feasible point works in the following way:

(i) A suitable $i = p \in I_2$ is selected. This may be, for example, the constraint which is the most difficult to satisfy.

(ii) Using the SUMT technique minimize the function

$$T(x, r) = -g_p(x) + r \sum_{i \in I_1} g_i^{-1}(x), \qquad (5.9.1)$$

until $g_p(x)$ becomes positive. If it happens that some other constraints $g_i(x)$, $i \in I_2$ are satisfied when $T(x, r)$ is minimized then they may be included in I_1.

(iii) If I_2 is nonempty then the procedure is repeated for a new $i \in I_2$ and the point obtained by solving (5.9.1). If min $[-g_p(x)] > 0$ for any selected p in the auxiliary problem, then it may indicate that the constraints of the original problem are inconsistent.

Other possibilities of finding an initial feasible point are discussed by Zoutendijk [10, 11].

Another problem we face is a proper choice of an initial value of the perturbation parameter $r = r_1$. This choice undoubtedly depends on the available starting point $x^{(0)}$. To illustrate possible difficulties that can occur, consider two extreme values of the initial perturbation parameter: (i) r_1 is very large, and (ii) r_1 is very small.

(i) If r_1 is too large then the term $\sum_i g_i^{-1}(x)$ is heavily weighted, and in minimizing $T(x, r_1)$ we can obtain a point $x(r_1)$ that is remote from the desired minimum. If in addition $x^{(0)}$ is a good approximation to the solution, then it is quite possible that $f(x(r_1)) > f(x^{(0)})$, which would be contrary to our intention of minimizing the object function $f(x)$. Figure 5.9.1 illustrates such a case when we consider our simple numerical problem (5.4.10): minimize $x_1 + x_2$ subject to the constraint $-x_1^2 - x_2^2 + 4x_1 + 4x_2 - 7 \geqslant 0$.

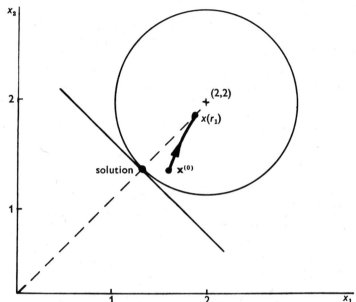

Figure 5.9.1. Illustration of the locations of $x^{(0)}$ and $x(r_1)$ if the perturbation parameter r is too large for problem (5.4.10).

(ii) Suppose that r_1 is very small and consider the problem: minimize $(x_1 - 1)^2 + (x_2 - 1)^2$ subject to the constraint $x_1 + x_2 - 1 = 0$. After the transformation we have

$$T(\mathbf{x}, r_1) = (x_1 - 1)^2 + (x_2 - 1)^2 + r_1^{-1/2}(x_1 + x_2 - 1)^2. \qquad (5.9.2)$$

For a small value of r_1 the penalty term involving equality is weighted heavily and this results in a very elongated and narrow valley near the solution which makes unconstrained minimization difficult (see Figures 5.9.2 and 5.9.3).

Similar difficulties can occur when problems including both equality and inequality constraints are considered. Thus far no sufficiently good formula for estimating r_1 has been developed, and the right choice of r_1 depends on the scaling of the problem.

For Problem I, Fiacco and McCormick [12] make the following suggestion: Suppose that $\mathbf{x}^{(0)}$ is given. A natural choice of r_1 seems to be that which minimizes $\|\nabla T(\mathbf{x}^{(0)}, r)\|$, i.e.,

$$\min_r \left(\nabla f(\mathbf{x}^{(0)}) - r \sum_i g_i^{-2}(\mathbf{x}^{(0)}) \nabla g_i(\mathbf{x}^{(0)}) \right)^2, \qquad (5.9.3)$$

which gives

$$r_1 = - \frac{(\nabla f(\mathbf{x}^{(0)}))^T \nabla \sum g_i^{-1}(\mathbf{x}^{(0)})}{(\nabla \sum g_i^{-1}(\mathbf{x}^{(0)}))^2}. \tag{5.9.4}$$

Formula (5.9.4) gives a reasonable value of r_1 provided that $\mathbf{x}^{(0)}$ is not too close to the boundary when the resulting value of r_1 can be too small. Rare trouble when $r_1 < 0$ may be overcome by moving along $-\nabla f(\mathbf{x}^{(0)})$ until r_1 is positive or an unconstrained minimum of $f(x)$ is found. In practice it has often proved adequate to start with a value of r_1 between 1 and 50, and to reduce this by a factor 10 at each stage.

The criterion for fixing the smallest value of r which is to be used in the sequence of T-function minimizations may be based on the results of the dual formulation. If a problem is convex we know that the exact optimum L_f is bounded by the primal and dual functions

$$f(\mathbf{x}^{(k)}) - r_k \sum_i g_i^{-1}(\mathbf{x}^{(k)}) \leqslant L_f \leqslant f(\mathbf{x}^{(k)}), \tag{5.9.5}$$

which leads to the simple stopping criterion

$$r_k \sum_i g_i^{-1}(\mathbf{x}^{(k)}) < \delta, \tag{5.9.6}$$

where

$$\delta \geqslant f(\mathbf{x}^{(k)}) - L_f > 0. \tag{5.9.7}$$

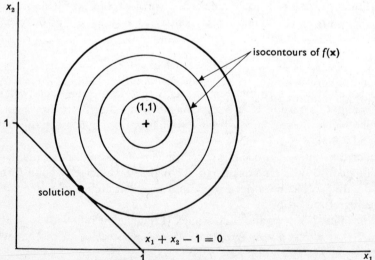

Figure 5.9.2. Isocontours of the function $f(\mathbf{x}) = (x_1 - 1) + (x_2 - 1)^2$ before the transformation $T(\mathbf{x}, r)$ is applied.

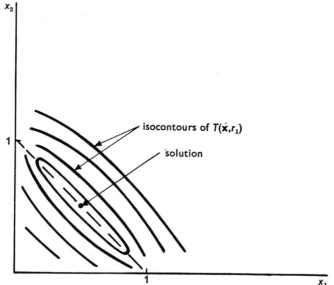

Figure 5.9.3. Isocontours of $T(x, r_1)$ (Problem 5.9.2) for a small value of r_1.

Other possible alternatives are

$$T(\mathbf{x}(r_{k-1}), r_{k-1}) - T(\mathbf{x}(r_k), r_k) < \delta \qquad (5.9.8)$$

or

$$f(\mathbf{x}(r_{k-1})) - f(\mathbf{x}(r_k)) < \delta, \qquad (5.9.9)$$

where δ is a preset positive member.

REFERENCES

1. M. J. Box, A Comparison of Several Current Optimization Methods and the Use of Transformations in Constrained Problems, *Computer J.*, 8(1966), 67.
2. R. Courant, Variational Methods for the Solution of Problems of Equililbrium and Vibrations, *Bull. Am. Math. Soc.*, 48(1943).
3. L. A. Schmit and R. L. Fox, *Advances in the Integrated Approach to Structural Synthesis*, A.I.A.A. Sixth Annual Structure and Materials Conference, Palms Springs, California, 1965.
4. C. W. Carroll, The Created Response Surface Technique for Optimizing Nonlinear, Restrained Systems, *Operations Res.*, 9(1961), 169.
5. A. V. Fiacco and G. P. McCormick, *Programming under Nonlinear Constraints by Unconstrained Minimization: A Primal-dual Method*, Research Analysis Corporation, RAC-TP-96, Bethesda, Md., 1963.

6. A. V. Fiacco and G. P. McCormick, *The sequential Unconstrained Minimization Technique for Convex Programming with Equality Constraints*, Research Analysis Corporation, RAC-TP-155, Bethesda, Md., 1965.
7. A. V. Fiacco and G. P. McCormick, Extensions of SUMT for Nonlinear Programming: Equality Constraints and Extrapolation, *Management Sci.*, *12*(1966), 816.
8. A. V. Fiacco and G. P. McCormick, *SUMT without Parameters*, Northwestern University, Evanson, Illinois, Syst. Res. Memorandum No. 121, 1965.
9. A. V. Fiacco, Comments on the Paper of C. W. Carroll, *Operations Res.*, *9*(1961), 184.
10. G. Zoutendijk, *Methods of Feasible Directions*, Elsevier Publishing Co., Amsterdam, and New York 1960, p. 66.
11. G. Zoutendijk, Nonlinear Programming: A Numerical Survey, *J. Soc. Indust. Appl. Math. Ser. A Control*, *4*(1966), 194.
12. A. V. Fiacco and G. P. McCormick, Computational Algorithm for the Sequential Unconstrained Minimization Technique for Nonlinear Programming. *Management Sci.*, *10*(1964), 601

Chapter 6

NUMERICAL RESULTS

6.1. INTRODUCTION

In this chapter we present the results of numerical experiments made using a selection from among the methods described in the preceding chapters. The plan of this chapter is as follows. We summarize the test problems in Section 2, consider some questions relating to the problem of scaling in Section 3, and describe the techniques used for minimization along a line in Section 4. The major comparisons are broken up in the same way as the chapters in which the relevant methods are described. Thus we consider direct search methods in Section 5, conjugate gradient methods in Section 6, and methods for minimizing sums of squares in Section 7. In Section 8 we illustrate the use of several of the minimization techniques for solving the sequence of unconstrained problems generated by the SUMT transformation.

In tabulating the results we have generally estimated the work required by giving the number of function calls. In methods which use derivatives it should be noted that the function value and the partial derivatives with respect to each independent variable are evaluated at each function call, and in several published comparisons of available techniques (for example Box [1]) what is given is the number of function calls multiplied by $(n + 1)$ where n is the number of independent variables. We have not followed this practice, and we suggest that the reader weigh the number of function calls using his knowledge of the function involved. We feel this is preferable to assuming that the calculation of each partial derivative can be equated with the function calculation.

The calculations have been carried out in single precision arithmetic on an IBM 360/50 computer. The programs are written in FORTRAN and PL/1.

6.2. THE SAMPLE PROBLEMS

Here we summarize briefly the problems on which the optimization techniques have been tested. Most have been taken from the literature but the Enzyme and Watson functions are new. These turn out to be somewhat more difficult than the other problems (a version of the Watson problem with 9 variables proving particularly difficult), and could prove useful additions to the comparatively limited number of test problems currently available.

(i) *Rosenbrock* [2]:

$$f = 100(x_2 - x_1^2)^2 + (1 - x_1)^2. \qquad (6.2.1)$$

This is perhaps the most widely used test function. Here f has a minimum of 0 at $x = (1, 1)$. There is a steep valley along the parabola $x_2 = x_1^2$.

(ii) *Cube* [3]:

$$f = 100(x_2 - x_1^3)^2 + (1 - x_1)^2. \qquad (6.2.2)$$

This is a variant on the Rosenbrock function, and has similar properties.

(iii) *Beale* [4]:

$$f = \sum_{i=1}^{3} (C_i - x_1(1 - x_2^i))^2, \qquad (6.2.3)$$

where $C_1 = 1.5$, $C_2 = 2.25$, $C_3 = 2.625$. This function has a narrow curving valley approaching $x_2 = 1$, and has minimum 0 at $x = (3, .5)$.

(iv) *Box* [1]:

$$f = \sum_{y} \{(e^{-x_1 y} - e^{-x_2 y}) - x_3(e^{-y} - e^{-10y})\}^2, \qquad (6.2.4)$$

where $y = 0.1(.1)1$. Box has described this function and the two-dimensional one obtained by setting $x_3 = 1$. In the latter case a contour plot shows a strongly asymmetric curved valley. Here the desired minimum is $f = 0$ at $x = (1, 10, 1)$.

(v) *Enzyme* [5]:

$$f = \sum_{i=1}^{11} \left(V_i - \frac{x_1(y_i^2 + x_2 y_i)}{y_i^2 + x_3 y_i + x_4}\right)^2. \qquad (6.2.5)$$

Table 6.2.1 gives the values of V_i, y_i, $x = 1, 2, \ldots, 11$. The minimum is $f = 3.075 \times 10^{-4}$ at $x = (.1928, .1916, .1234, .1362)$. The origin of this problem has been described in detail by Kowalik and Morrison [6].

(vi) *Watson*:

$$f = \sum_{i=1}^{30} \left\{\sum_{j=1}^{m}(j - 1)x_j y_i^{j-2} - \left(\sum_{j=1}^{m} x_j y_i^{j-1}\right)^2 - 1\right\}^2 + x_1^2, \qquad (6.2.6)$$

where $y_i = (i - 1)/29$. In this problem an attempt is made to approximate to the solution of the differential equation

$$\frac{dz}{dx} - z^2 = 1, \qquad z(0) = 0, \qquad (6.2.7)$$

Table 6.2.1

i	V_i	y_i
1	.1957	4
2	.1947	2
3	.1735	1
4	.1600	.5
5	.0844	.25
6	.0627	.167
7	.0456	.125
8	.0342	.1
9	.0323	.0823
10	.0235	.0714
11	.0246	.0625

in $0 \leqslant x \leqslant 1$ by a polynomial of degree m by minimizing the sum of squares of the residuals at selected points. For the points indicated, the solution for $m = 6$ is $f = 2.288 \times 10^{-3}$ at $\mathbf{x} = (-.016, 1.012, -.233, 1.260, -1.513, .993)$. This is not an example for which polynomial approximation is suitable, so that the comparative difficulty of this problem could have been anticipated.

(vii) *Rosen-Suzuki* [7]:

$$f = x_1^2 + x_2^2 + 2x_3^2 + x_4^2 - 5x_1 - 5x_2 - 21x_3 + 7x_4 \qquad (6.2.8)$$

subject to

$$-x_1^2 - x_2^2 - x_3^2 - x_4^2 - x_1 + x_2 - x_3 + x_4 + 8 \geqslant 0, \qquad (6.2.9)$$
$$-x_1^2 - 2x_2^2 - x_3^2 - 2x_4^2 + x_1 + x_4 + 10 \geqslant 0, \qquad (6.2.10)$$
$$-2x_1^2 - x_2^2 - x_3^2 - 2x_1 + x_2 + x_4 + 5 \geqslant 0. \qquad (6.2.11)$$

This function has a minimum $f = -44$ at $\mathbf{x} = (0, 1, 2, -1)$. The constraint (6.2.11) is active.

(viii) *Beale* [8]:

$$f = 9 - 8x_1 - 6x_2 - 4x_3 + 2x_1^2 + 2x_2^2 + x_3^2 + 2x_1x_2 + 2x_1x_3 \qquad (6.2.12)$$

subject to

$$x_1 \geqslant 0, \qquad x_2 \geqslant 0, \qquad x_3 \geqslant 0 \qquad (6.2.13)$$

and

$$x_1 + x_2 + 2x_3 \leqslant 3. \qquad (6.2.14)$$

The solution is $f = \frac{1}{9}$ at $\mathbf{x} = (\frac{4}{3}, \frac{7}{9}, \frac{4}{9})$, and the constraint (6.2.14) is active.

6.3. SCALING AND RELATED PROBLEMS

The scaling of the variables in an optimization problem is often crucial to the successful functioning of existing methods. Intuitively we mean by "well-scaled" that similar changes in the variables lead to similar changes in the objective function. In a contour map of a well-scaled function the contour lines would not deviate too far from concentric circles, and in this case we would hope that the method of steepest descent would work satisfactorily. When the contours are exactly concentric circles steepest descent gives the answer in one step.

In general it does not appear to be easy to impose a satisfactory scaling on the variables in a problem. However, gross inequalities should be removed as a matter of course.

Although little can be done in the large, it is possible to scale the function in the neighborhood of the minimum on the basis of our standard assumption that it is here represented adequately by a quadratic form. Such a scaling is relevant when considering stopping criteria.

We have

$$F = \tfrac{1}{2}\mathbf{x}^T A \mathbf{x}$$

$$= \tfrac{1}{2}\mathbf{x}^T Q^T \Lambda Q \mathbf{x}$$

$$= \tfrac{1}{2}\boldsymbol{\xi}^T \boldsymbol{\xi} \tag{6.3.1}$$

where

$$\boldsymbol{\xi} = \sqrt{\Lambda}\, Q \mathbf{x} \tag{6.3.2}$$

We have assumed that $A = Q^T \Lambda Q$ where Q is orthogonal, and Λ is the diagonal matrix of the eigenvalues which are positive as $\mathbf{x} = 0$ minimizes F. The curves $F = $ constant are circles in the new variables.

The interesting term in the transformation is $\sqrt{\Lambda}$, and to explain its significance it is convenient to introduce a new variable $\mathbf{z}$ given by

$$\mathbf{z} = Q\mathbf{x} \tag{6.3.3}$$

which is a change of variables to the principal axes of the quadratic form. Now the ellipsoid $F = $ constant cuts the i-th principal axis at the points $z_i = \pm\sqrt{F/\lambda_i}$, so that small λ_i correspond to elongation of the ellipsoid in the direction of the corresponding axis, and this implies in turn that the corresponding variable is poorly determined in the optimization calculation. Because of the term $\sqrt{\Lambda}$ in equation (6.3.2) the poorly determined variables have small weights after the transformation, and this is intuitively desirable.

In ξ space where the variables are in a sense optimally scaled the distance of points on $F = \text{constant}$ from the origin (which is also the solution) is given by

$$\|\xi\| = \sqrt{2F} \tag{6.3.4}$$

and this provides a suitable measure of the absolute error in the current approximation when $F = 0$ at the minimum.

When F is not zero at the minimum and is therefore unknown then a somewhat more complicated estimate is required. Let ξ_i and ξ_{i+1} be two approximations to the solution. Then the results of Section 3.8 show that provided $(\mathbf{x}_{i+1} - \mathbf{x}_i)^T \mathbf{g}_{i+1} = 0$, then

$$\|\xi_{i+1} - \xi_i\| = \sqrt{\{(\mathbf{x}_{i+1} - \mathbf{x}_i)^T A (\mathbf{x}_{i+1} - \mathbf{x}_i)\}}$$
$$= \sqrt{2(F(\mathbf{x}_i) - F(\mathbf{x}_{i+1}))} \tag{6.3.5}$$

As $\|\xi_{i+1} - \xi_i\| < \|\xi_{i+1}\| + \|\xi_i\|$ equation (6.3.5) can underestimate the distance of ξ_{i+1} and ξ_i from the solution. However there are two ways in which it can be used effectively:

1. We can argue that it is reasonable to expect at least linear convergence for a general descent calculation. That is, we anticipate that an inequality of the form

$$\|\xi_{i+1} - \xi_i\| \leqslant \gamma \|\xi_i - \xi_{i-1}\| \tag{6.3.6}$$

holds for $i = 1, 2, \ldots$, where $\gamma < 1$. Then we have

$$\|\xi_p\| \leqslant \|\xi_p - \xi_{p+1}\| + \|\xi_{p+1} - \xi_{p+2}\| + \cdots$$
$$\leqslant \frac{1}{1 - \gamma} \|\xi_p - \xi_{p+1}\|$$
$$\leqslant \frac{1}{1 - \gamma} \sqrt{2(F(\mathbf{x}_p) - F(\mathbf{x}_{p+1}))}, \tag{6.3.7}$$

which should provide a useful stopping criterion. Presumably estimates of γ would be calculated as the computation proceeds.

2. If the rate of convergence is sufficiently rapid (so that $\|\xi_{i+1}\| \ll \|\xi_i\|$) then equation (6.3.5) could be applied as it stands. A possible application is to the conjugate gradient methods where ξ_i and ξ_{i+1} are points reached after successive sweeps through n conjugate directions.

In practice it seems at least as difficult to prevent computations terminating before the solution has been reached, and quite elaborate techniques to guard against this have been suggested in the literature. For example, Powell

[9] has suggested that when a tentative minimum has been calculated this point should be suitably perturbed and the calculation restarted from the perturbed point. If the same minimum results this is then accepted. However, if some other point is found then the line joining them is used to define the direction of a linear search, and from the point reached in this search the computation is again restarted.

Another application of the technique of local scaling that suggests itself is to expand the function to be minimized in a Taylor series up to terms of second order, make the transformation to the ξ variables, and then apply a step of the method of steepest descent to the transformed problem. It is interesting that the resulting method is equivalent to an application of a step of Newton's method to the system of equations $\nabla F(\mathbf{x}) = 0$, and this possibly explains why this method often requires substantially fewer iterations for convergence than other methods (although requiring a substantial amount of work in each iteration). In deriving the Davidon method we pointed out that it could be considered as an attempt to perform a Newton step using the best available approximation to the inverse, so that it seems likely that the generally good performance of the Davidon method can be attributed to comparative scale independence.

6.4. MINIMIZING A FUNCTION ON A LINE

Most of the techniques we have considered are based on the assumption that the minimum of a function of a single variable is readily calculated. We summarize here the methods used in this important operation in obtaining the results reported elsewhere in this chapter.

Three types of technique have been used corresponding to three fairly distinct applications:

(i) *Descent Methods Making Use of Derivatives*. Here we have the ALGOL program given by Fletcher and Reeves [10]. In this case the derivative along the line is available in addition to the function value so that only two function calls are required to bracket the minimum, and this also gives sufficient information to determine a cubic interpolation polynomial.

(ii) *Descent Methods Using Function Values Only*. In this case a modified form of the second scheme for repeated quadratic interpolation (Chapter 2, Section 3) has been used. In this the quadratic is fitted to the (usually non-equispaced) points making up the current bracket. If the minimum of the quadratic lies within the bracket it replaces one of the original points and the procedure is repeated. Otherwise a point lying in the range of the bracket is chosen as the average of two of the points making up the bracket.

(iii) *Calculations Associated with the SUMT Transformation.* Here a routine based on the golden section algorithm is employed. Basically this is because we have had previous experience of the satisfactory performance of this algorithm in this application [11], and because it is a comparatively easy task with this algorithm to take account of any violation of constraints. Presumably the other techniques could be adapted to this purpose almost as readily, but there is the possibility that quadratic interpolation may not be effective as the optimum approaches the active constraints.

The performance of the linear search routine is affected, in particular, by the choice of the parameters governing the size of the initial interval in the search for a bracket, and the tolerance to which the minimum is to be found.

For the second of our techniques the influence of both these factors has been carefully evaluated for the particular case of the Powell sum of squares method (Chapter 4, Section 12) applied to the Rosenbrock function. The tests were as follows:

(a) Comparison of different search intervals for the initial minimization along a line (subsequently the search interval was taken as the displacement in the previous step of the iteration). Here we noted a remarkable variation of up to 40% in the number of function evaluations needed, and a similar variation in the number of minimization steps. The optimum interval was of the order of $\frac{1}{10}$ of the displacement from the starting point $(-1.2, 1)$ to the solution. It was not a good strategy to take a much smaller interval.

(b) Comparison of the strategy of taking the displacement in the previous step of the iteration as the initial search interval for the next linear search against a strategy in which the search interval is less liable to fluctuations (the average of the displacements in all previous steps was tried). Here the strategy of taking the displacement in the previous step was clearly superior in the final approach to the minimum (for the overall problem). Possibly the averaging technique did not allow the search interval to shrink rapidly enough at this stage. Earlier there was little to choose between the strategies.

(c) Comparison of the effect of two different tolerances and a strategy in which the routine is left as soon as the quadratic gives a point with a function value lower than any previously recorded. In this case the third strategy proved vastly superior. For example it took only about $\frac{1}{3}$ of the number of function evaluations and (surprisingly) little more than $\frac{1}{2}$ the number of linear searches that were required when the tolerance on each linear search was the same as that on the overall problem. A less restrictive criterion in which the minimum was required to be located in an interval $\frac{1}{10}$ the length of the initial bracket gave a somewhat improved performance. However, the third strategy was still markedly superior.

We remark that these results are probably highly problem dependent and, possibly, are also technique dependent as well. The sensitivity to variations in their input data of several of the methods we consider here is remarked on subsequently, and indicates a less than satisfactory state of affairs.

In the Newton and Davidon methods, because of the local scaling that is effected behind the scenes, the choice of a unit initial interval is appropriate in attempting to find a bracket.

6.5.   DIRECT SEARCH METHODS

In this section we compare the numerical results produced by the Hooke and Jeeves, Rosenbrock, and Simplex methods. All three methods have shown a comparable efficiency when tested on problems with a small number of independent variables (Tables 6.5.1 and 6.5.3). However, for the Watson and Enzyme functions (Tables 6.5.4 to 6.5.6) the Simplex method has shown superiority over the other two techniques. The Watson function proves to be the most difficult for all three methods, the Simplex method converging successfully after 900 function evaluations, while the same number of function evaluations is not sufficient for the Rosenbrock method to produce comparable accuracy. The method of Hooke and Jeeves has shown very slow convergence and has not been tabulated for this reason.

It is very interesting that our experimental results show that the efficiency of the Simplex method is comparable with that of the Powell conjugate gradient method. Our experience therefore supports the results claimed for the Simplex method by Nelder and Mead [12]. Box [1] has remarked that the superior performance of the Simplex method does not continue as the dimensionality of the problem increases. We cannot offer a definite conclusion, and suggest that the relative performance of these two methods could be decided only after sufficient experimental work is done. At present such numerical evidence is not available.

It is essential to note that the efficiency of the Rosenbrock and Simplex methods depends on the choice of the coefficients α, β, γ, and, in the latter case, on the size of the initial regular simplex. To illustrate the variation in performance of the Rosenbrock method we have summarized in Table 5.6.7 the results of applying it to the Cubic function using four different sets of coefficients α, β. It is clear that the best choice of α, β depends on the particular function. However, $\alpha = 5$, $\beta = .5$ and $\alpha = 3$, $\beta = .5$ have been satisfactory for all our test functions.

Table 6.5.8 illustrates the influence of the initial size of the regular simplex on the efficiency of the Simplex method. This and other results indicate that

Table 6.5.1. Comparison of Direct Search Methods

No. of funct. eval.	Hooke and Jeeves			Rosenbrock, $\alpha = 5$, $\beta = .5$			Simplex, $\alpha = 1$, $\beta = .5$, $\gamma = 2$		
	x_1	x_2	f	x_1	x_2	f	x_1	x_2	f
50	−.9562	.9187	3.83	.5795	.3232	.19	−.2085	.0175	1.52
100	−.1499	.0123	.9352	.9352	.8729	4.48×10^{-3}	.5557	.3192	.21
200	.9438	.8936	3.92×10^{-3}	1.0016	1.0031	4.37×10^{-6}	1.0002	1.0004	6.6×10^{-8}
250	1.0001	1.0001	1.25×10^{-8}						

Rosenbrock, starting point $\mathbf{x}^{(0)} = (−1.2, 1)$, $f = 24.20$.

Table 6.5.2.   Comparison of Direct Search Methods

No. of funct. eval.	Hooke and Jeeves				Simplex, $\alpha = 1$, $\beta = .5$, $\gamma = 2$				$\alpha = 5$, $\beta = .5$ Rosenbrock
	x_1	x_2	x_3	f	x_1	x_2	x_3	f	
50	1	36	1	.14	.3912	26.2933	1.2800	.17	Failed to converge and produced the following result
100	1	10	1	0	.6137	26.8182	1.2905	.05	
290					1.0122	9.7460	.9900	3.8×10^{-5}	$x_1 \to \infty$ $x_2 \to \infty$ $x_3 \to 0$

Box, starting point $\mathbf{x}^{(0)} = (0, 20, 20)$, $f = 1021.67$.

Table 6.5.3. Comparison of Direct Search Methods

No. of funct. eval.	Hooke and Jeeves			Rosenbrock, $\alpha = 5, \beta = .5$			Simplex, $\alpha = 1, \beta = .5, \gamma = 2$		
	x_1	x_2	f	x_1	x_2	f	x_1	x_2	f
50	2.6999	.4000	2.42×10^{-2}	2.9512	.5019	4.88×10^{-3}	3.0550	.5122	4.87×10^{-4}
100	3	.5	0	2.9802	.4955	6.78×10^{-5}	2.9998	.4999	2.01×10^{-8}
130				2.9989	.49967	2.61×10^{-7}			

Beale, starting point $\mathbf{x}^{(0)} = (0, 0)$, $f = 14.20$.

Table 6.5.4. Comparison of Direct Search Methods

No. of funct. eval.	Hooke and Jeeves		Rosenbrock, $\alpha = 3$, $\beta = .5$		Simplex, $\alpha = 1$, $\beta = .5$, $\gamma = 2$	
	x	f	x	f	x	f
100	.2249		.1987		.1864	
	−.0437	9.71×10^{-4}	.0455	3.97×10^{-4}	.0662	6.21×10^{-4}
	.2375		.0898		−.0298	
	.0000		.0680		.0979	
400	.1953		.1931		.1927	
	.1644	3.10×10^{-4}	.1756	3.08×10^{-4}	.1918	3.07×10^{-4}
	.1264		.1163		.1228	
	.1234		.1293		.1364	
600	.1933					
	.1785	3.07×10^{-4}				
	.1201					
	.1302					

Enzyme, starting point $\mathbf{x}^{(0)} = (0, 0, 0, 0)$, $f = .1484$.

Table 6.5.5. Comparison of Direct Search Methods

No. of funct. eval.	Hooke and Jeeves		Rosenbrock, $\alpha = 3$, $\beta = .5$		Simplex, $\alpha = 1$, $\beta = .5$, $\gamma = 2$	
	x	f	x	f	x	f
100	.2187		.1902		.1808	
	.2462	8.53×10^{-4}	.4346	3.79×10^{-4}	.6936	4.34×10^{-4}
	.4337		.2675		.3369	
	.1337		.2300		.3304	
300	.1898		.1839		.1928	
	.2712	3.16×10^{-4}	.4111	3.53×10^{-4}	.1925	3.07×10^{-4}
	.1493		.1727		.1240	
	.1697		.2301		.1366	
500 R., 600 H&J	.1926		.1926		—	—
	.1982	3.07×10^{-4}	.1948	3.07×10^{-4}		
	.1257		.1234			
	.1391		.1376			

Enzyme, starting point $x^{(0)} = (.25, .4, .4, .4)$, $f = 5.56 \times 10^{-3}$.

Table 6.5.6. Comparison of Direct Search Methods

No. of funct. eval.	Rosenbrock		Simplex	
	x	f	x	f
300	-.079		-.045	
	1.004		.971	
	.038	1.53×10^{-2}	.200	1.07×10^{-2}
	-.049		-.175	
	.265		.201	
	.134		.256	
600	-.045		-.002	
	.971		1.019	
	.195	1.02×10^{-2}	-.209	3.29×10^{-3}
	-.141		1.242	
	.147		-1.534	
	.284		1.042	
900	-.041		-.015	
	.970		1.012	
	.200	9.64×10^{-3}	-.232	2.288×10^{-3}
	-.098		1.259	
	-.059		-1.513	
	.336		.993	

Watson, starting point $x^{(0)} = (0, 0, 0, 0, 0)$, $f = 30.00$.

Table 6.5.7.   Rosenbrock Method

No. of funct. eval.	$\alpha = 5, \beta = .5$		$\alpha = 3, \beta = .5$		$\alpha = 1.5, \beta = .5$		$\alpha = 2, \beta = .3$	
	x	f	x	f	x	f	x	f
100	.8251 .5607	3.06×10^{-2}	1.0024 1.0074	1.27×10^{-5}	.9115 .7542	8.84×10^{-3}	.7068 .3433	9.54×10^{-2}
200	1.0000 1.0000	0	1.0000 1.0000	0	.9408 .8320	3.56×10^{-3}	.7162 .3666	8.05×10^{-2}
300					1.0000 1.0000	0	.7298 .3867	7.34×10^{-2}

Cubic, starting point $\mathbf{x}^{(0)} = (-1.2, 1)$, $f = 749.03$.

Table 6.5.8. Simplex Method ($\alpha = 1$, $\beta = .5$, $\gamma = 2$)

No. of funct. eval.	Size of the initial regular simplex					
	1		.5		.1	
	x	f	x	f	x	f
50	.6642 .2947	1.13×10^{-1}	.8314 .5705	3.01×10^{-2}	.3025 .3459	4.91×10^{-1}
100	.9345 .8151	4.39×10^{-3}	.9678 .9054	1.14×10^{-3}	.7007 .3183	1.56×10^{-1}
140	.9998 .9997	1.56×10^{-6}	1.0000 .9999	1.3×10^{-7}	.8243 .5552	3.33×10^{-2}
206					.9999 .9999	3.79×10^{-8}

Cubic, starting point $\mathbf{x}^{(0)} = (-1. \ 2, \ 1)$, $f \doteq 749.03$.

it is easier to contract rather than expand the simplex, and for this reason it is important that the initial simplex be large enough. We have also concluded that the simple strategy $\alpha = 1$, $\beta = .5$, $\gamma = 2$ is a good and safe one, and this supports the Nelder and Mead recommendation.

6.6. DESCENT METHODS

In this section we report our experience in using the methods of Davidon, Fletcher and Reeves, and Powell (without derivatives). Our programs for the Davidon and Fletcher-Reeves methods correspond closely to published ALGOL procedures [10, 13]. However the following variation should be noted. Fletcher's program for Davidon's method (FLEPOMIN) tests the value of $-\mathbf{g}^T H \mathbf{g}$ before the minimization along the line is started. If this value is positive so that $\mathbf{d} = -H\mathbf{g}$ is not downhill then the procedure is terminated. In theory, H is positive-definite, but due to rounding errors it may occasionally lose this property. In such a case we restart the procedure from the current point with the initial positive-definite matrix. This simple device ensures that the computation is not stopped prematurely.

The Powell conjugate gradient method has been programmed following the suggestions given by Powell [9] and is presented here in Chapter 3, Section 8. Powell has reported that the simple form of his algorithm is unsatisfactory, and it occurred to us that a restart procedure similar to that used in successful implementation of the Fletcher-Reeves method could be used. It is interesting that this was noticably inferior to a program based on the Powell recommendations when applied to the Rosenbrock function.

It is clear from the numerical results collected in Tables 6.6.1 to 6.6.7 that the comparison between the Davidon and Fletcher and Reeves methods favors the former. This result has been reported by other investigators [1, 10], who have offered the following explanation. The Davidon procedure continuously collects the information about the actual surface, at the price of the H matrix storage and rather complex computations. In contrast the Fletcher and Reeves method restarts the process periodically from the best point yet obtained and in so doing throws away all the information accumulated in the previous stage of the iteration. However, as we have mentioned in Chapter 3, Section 7, the Fletcher and Reeves method might be preferred in problems in which the number of variables is so large that the H matrix must be held in the backing store.

It will be noted that if the conventional $(n + 1)$ weighting factor is applied to the number of function calls in the Davidon method then the Powell

Table 6.6.1. Comparison of Descent Methods

No. of funct. calls	Davidon		Fletcher and Reeves		Powell	
	x	f	x	f	x	f
50	.6885 .4580	1.22×10^{-1}	.5093 .2132	4.52×10^{-1}	-.4769 .1884	2.33
80	.9992 .9984	6.27×10^{-7}	1.0035 1.0068	2.16×10^{-5}	.1198 .0224	9.09×10^{-1}
200					.9999 .9999	5.57×10^{-7}

Rosenbrock, starting point $\mathbf{x}^{(0)} = (-1.2, 1)$, $f = 24.20$.

Table 6.6.2. Comparison of Descent Methods

No. of funct. calls	Davidon		Fletcher and Reeves		Powell	
	x	f	x	f	x	f
50	.9421 .8334	4.16×10^{-3}	1.0000 1.0000	0	.9340 .8114	5.46×10^{-3}
75	1.0000 1.0000	0			.9769 .9300	9.68×10^{-4}
110					1.0000 1.0000	0

Cubic, starting point $\mathbf{x}^{(0)} = (-1.2, 1)$, $f = 749.04$.

Table 6.6.3. Comparison of Descent Methods

No. of funct. calls	Davidon		Fletcher and Reeves		Powell	
	x	f	x	f	x	f
20	2.9999 .5000	2.18×10^{-11}	2.7864 .4532	5.72×10^{-3}	2.4984 .2723	.136
70			2.9985 .4996	2.03×10^{-7}	2.9972 .4988	6.37×10^{-6}
86					2.9998 .4999	2.94×10^{-8}

Beale, starting point $\mathbf{x}^{(0)} = (.1, .1), f = 6.49.$

Table 6.6.4. Comparison of Descent Methods

No. of funct. calls.	Davidon		Fletcher and Reeves		Powell	
	x	f	x	f	x	f
30	1.000 10.000 1.000	0	.966 10.419 1.024	8.63×10^{-5}	Has not been recorded	
70			1.000 10.001 1.000	3.07×10^{-9}	.435 22.106 1.377	6.66×10^{-2}
205					.999 10.000 1.000	3.09×10^{-9}

Box, starting point $\mathbf{x}^{(0)} = (0, 10, 20)$, $f = 1031.2$.

Table 6.6.5. Comparison of Descent Methods

No. of funct. calls	Davidon		Fletcher and Reeves		Powell	
	x	f	x	f	x	f
50	.1852		.1927		.1545	
	.0928	4.76×10^{-4}	.0003	7.64×10^{-4}	$-.0634$	7.86×10^{-3}
	.0088		.0022		.0233	
	.1003		.0570		.0069	
110	.1928		.2106		.2124	
	.1916	3.07×10^{-4}	.0218	4.91×10^{-4}	$-.0519$	1.83×10^{-3}
	.1234		.1728		.1526	
	.1362		.0463		.0199	
405 F&R, 425 P			.1931		.1928	
			.1958	3.07×10^{-4}	.1910	3.07×10^{-4}
			.1285		.1230	
			.1378		.1359	

Enzyme, starting point $\mathbf{x}^{(0)} = (0, 0, 0, 0)$, $f = .1484$.

Table 6.6.6. Comparison of Descent Methods

No. of funct. calls	Davidon		Fletcher and Reeves		Powell	
	x	f	x	f	x	f
30	.1857 .2966 .1211 .1821	3.31×10^{-4}	.1944 .5183 .3572 .2629	4.54×10^{-4}	.2074 .7033 .6172 .3497	7.97×10^{-4}
60	.1928 .1912 .1230 .1360	3.07×10^{-4}	.1844 .5424 .2687 .2777	3.9×10^{-4}	.2013 .7463 .6257 .3482	6.91×10^{-4}
90			.1929 .1834 .1189 .1326	3.07×10^{-4}	.1928 .1909 .1229 .1359	3.07×10^{-4}

Enzyme, starting point $x^{(0)} = (.25, .39, .415, .39)$, $f = 5.31 \times 10^{-3}$.

Table 6.6.7. Comparison of Descent Methods

No. of funct. calls	Davidon x	Davidon f	Fletcher and Reeves x	Fletcher and Reeves f	Powell x	Powell f
65 D, F&R 75 P	$-.016$ 1.012 $-.232$ 1.262 -1.513 .992	2.287×10^{-3}	$-.064$ .977 .130 $-.021$ .045 .288	9.815×10^{-3}	.023 .769 .244 .206 .178 .064	4.616×10^{-1}
300			$-.038$ .980 .105 .156 .205 .434	7.180×10^{-3}	.078 .985 .121 .135 .256 .169	1.276×10^{-2}
700			$-.037$ .985 .061 .289 $-.358$ .496	6.077×10^{-3}	$-.019$ 1.011 $-.198$ 1.152 1.395 .945	2.434×10^{-3}
815 P, 1700 F&R			$-.022$ 1.004 $-.443$ .961 -1.156 .837	2.643×10^{-3}	$-.019$ 1.012 $-.321$ 1.254 -1.506 .989	2.288×10^{-3}

Watson, starting point $\mathbf{x}^{(0)} = (0, 0, 0, 0, 0)$, $f = 30$.

method actually compares favorably with it. However, our experience has been that the Davidon method is the most likely of all to work and it is the method we would try first if derivatives are readily available.

6.7. MINIMIZING A SUM OF SQUARES

The major algorithms for nonlinear problems discussed in Chapter 4 are Newton's method, the secant algorithm, the Gauss method, the algorithm of Levenberg, Marquardt, and Morrison, and Powell's method. Among these methods the secant algorithm and Powell's method are of particular interest as they do not require derivatives, and our attention has been largely restricted to them. Comparisons between the Gauss method and that of Powell have been given by Box [1], and detailed information on the thorough testing of an implementation of the Marquardt algorithm can be found in the SHARE Secretary Distribution 164, 11th April 1967.* This article gives an interesting account of the problems involved in using a computer library subroutine for optimization problems.

Although their application is limited to objective functions of a special form, the secant and Powell methods are among the most powerful currently available. The principal problem areas in implementing them concern (i) the provision of an appropriate to matrix in starting the computation, and (ii) the selection of the column to be replaced in updating the G matrix after the current step of the iteration.

The similarity between the two methods has already been stressed, and similar performance can be anticipated from them in problems to which both are applicable.

We have tried two starting procedures. In the first we compute a single relaxation sweep, and after finding the minimum in each search along a coordinate direction the appropriate derivatives are estimated by differencing function values used in the linear search. This method requires more function values than does the second, which involves no searching. In the second method function values are used only in estimating derivatives by differences.

Tests of these starting procedures used with the Powell algorithm have been run using the Rosenbrock and Enzyme problems and an easy nine variable problem coming from finite difference approximation to a nonlinear differential equation. The most significant results appear to come from the Enzyme problem where the relaxation start was markedly superior. This is illustrated in Table 6.7.1. It should be noted that both calculations compare more than favorably with those previously reported.

* Periodic publication available from the secretary or any member of SHARE, the users' group for IBM 7090 and System 360 computers.

Table 6.7.1

Starting procedure	Numbers of iterations	Number of function calls
1	17	97
2	33	191

Enzyme (Powell method), starting point $\mathbf{x}^{(0)} = (.25, .39, .415, .39)$, $f = 5.31 \times 10^{-3}$.

Further experiments with the Rosenbrock function do not confirm the difference, and both starting procedures work well for this problem. However, with the easy nine variable problem the second starting procedure is very good. It effectively gives the answer in one iteration while four iterations are needed to reach the same point after the relaxation start.

It should be noted that for large problems for which few iterations are needed the second start can be markedly superior because the relaxation start can require many more linear searches than does the subsequent solution. In this case the use of the second start can reduce problem time by a factor of the order $\frac{1}{3}$ or better. This has been well borne out using the secant method on a problem with 75 variables.

The second problem in implementing these methods concerns the selection of the column of G to be replaced. Here two alternatives have been discussed. These are the elaborate criterion due to Powell which is described in Chapter 4, Section 12, and the simpler procedure used by Barnes of replacing the columns in sequence. Here we tested both alternatives using the Powell algorithm applied to the Enzyme problem. A relaxation start was used in both cases. In this case the Powell criterion gave the results displayed in the first row of Table 6.7.1, while the sequential technique failed to give satisfactory convergence and the desired minimum was not reached. This result would appear to be a good guide to the value of Powell's criterion. We mention that our version of the secant algorithm used the sequential replacement technique. This is equivalent to that of Powell for problems with two variables (such as the Rosenbrock function). The only tests using more variables have been on functions that turned out to be comparatively easy, and the performance of the algorithm has been very satisfactory. However more detailed testing is obviously desirable.

6.8. APPLICATIONS OF THE SUMT TRANSFORMATION

The SUMT transformation described in Chapter 5 is interesting not only because it provides a method by which a range of important practical

Table 6.8.1. Davidon Method and SUMT Transformation

r	x_1	x_2	x_3	x_4	$T(\mathbf{x}, r)$	$f(\mathbf{x})$	No. of funct. calls
1	$-.0066$	.9068	1.8995	$-.8225$	-38.62	-41.43	43
10^{-2}	$-.0128$	.9962	1.9939	$-.9834$	-43.51	-43.76	85
10^{-4}	$-.0008$	.9993	1.9991	$-.9987$	-43.95	-43.98	125
10^{-6}	.0003	.9997	2.0001	$-.9997$	$-43.99(5)$	-44	160

Rosen-Suzuki, starting point $\mathbf{x}^{(0)} = (0, 0, 0, 0)$, $f = .425$.

Table 6.8.2. Davidon Method and SUMT Transformation

r	x_1	x_2	x_3	$T(x, r)$	$-f(x)$	No. of funct. calls
1	.8946	.7046	.4288	7.4157	.7055	80
10^{-2}	1.3798	.7362	.3506	.2598	.1557	180
10^{-4}	1.3605	.7706	.4231	.1215	.1164	240
10^{-5}	1.3355	.7763	.4407	.141	.1126	Approx. 300

Beale, starting point $x^{(0)} = (.1, .1, .1)$, $f = 7.29$.

Table 6.8.3. DSC Method and SUMT Transformation

r	x_1	x_2	x_3	$T(\mathbf{x}, r)$	$f(\mathbf{x})$	No. of funct. calls
1	.8884	.7188	.7260	7.4168	.7072	210
10^{-2}	1.3313	.7539	.3710	.2605	.1564	330
10^{-4}	1.3478	.7720	.4293	.1210	.1158	510
10^{-6}	1.3620	.7687	.4335	.1128	.1124	560

Beale, starting point $\mathbf{x}^{(0)} = (.1, .1, .1)$, $f = 7.29$.

Table 6.8.4. Powell Method and SUMT Transformation

r	x_1	x_2	x_3	$T(\mathbf{x}, r)$	$f(\mathbf{x})$	No. of funct. calls
1	.8944	.7117	.7337	7.4118	.6896	195
10^{-2}	1.3944	.7237	.3528	.2602	.1541	340
10^{-4}	1.3394	.7735	.4326	.1210	.1160	530
10^{-6}	1.3402	.7835	.4366	.1123	.1119	650

Beale, starting point $\mathbf{x}^{(0)} = (.1, .1, .1, .1)$, $f = 7.29$.

problems may be tackled but also because the transformed problems can be expected to be severe tests of the methods applied to solve them. The reason for this is that if the constrained problem has its solution on the boundary of the feasible region then the convergence of the transformed objective functions will not, in general, be uniform. Also, as the active constraint is approached the assumption that the objective function can be adequately approximated by a quadratic form in a neighborhood of the minimum will at best be true only in a very small neighborhood. However, there is also the advantage that the minimum for $r = r_k$ will be a very good approximation to the minimum for $r = r_{k+1}$ as r becomes small.

We can argue that conjugate direction methods should work well because when r is not small they should function normally, and when $r \to 0$ they should still work as descent methods applied to find a solution from a very good first approximation. In our experiments we have tried the Davidon and Powell (conjugate direction) methods, and we have also used the Davies, Swann, and Campey (or DSC) method which we described in outline in Chapter 3, Section 8. This is a descent method that does not use conjugate directions. We give in Table 6.8.1 the results of applying the Davidon method to the Rosen-Suzuki function, and we note that similar results have been obtained with the other methods. The final number of function calls for both of them was 900. The results for the Beale functions are given in Tables 6.8.2 to 6.8.4. We note the good performance of the DSC method. This would tend to confirm our suggestion that ultimately it is the descent nature of the computation rather than any special property of conjugate directions that is important here.

These problems do little more than indicate the feasibility of the SUMT approach to mathematical programming. However there is a growing literature of successful applications [14].

REFERENCES

1. M. J. Box, A comparison of Several Current Optimization Methods and the Use of Transformations in Constrained Problems, *Computer J.*, 9(1966), 67.
2. H. H. Rosenbrock, An Automatic Method for Finding the Greatest or Least Value of a Function, *Computer J.*, 3(1960), 175.
3. A. Leon, A Comparison among Eight Known Optimizing Procedures, in *Recent Advances in Optimization Techniques*, A. Lavi and T. P. Vogl, eds., John Wiley & Sons, New York, 1960, p. 33.
4. E. M. L. Beale, On an Iterative Method for Finding a Local Minimum of a Function of More Than One Variable, *Statistical Techniques Research Group*, Princeton University, Tech. Report 25, 1958.

5. J. Kowalik, A Note on Nonlinear Regression Analysis, *Austral. Computer J.*, *1*(1967), 51.

6. J. Kowalik and J. F. Morrison, Analysis of Kinetic Data for Allosteric Enzyme Reactions as a Nonlinear Regression Problem, *Math. Biosciences*, *2*(1968), 57–66.

7. J. B. Rosen and S. Suzuki, Construction of Nonlinear Programming Test Problems, *Comm. ACM*, *8*(1965), 113.

8. E. M. L. Beale, Numerical Methods, in *Nonlinear Programming*, *J. Abadie*, ed., North-Holland Publishing Co., Amsterdam, 1967, p. 150.

9. M. J. D. Powell, An Efficient Method for Finding the Minimum of a Function without Calculating Derivatives, *Computer J.*, *7*(1964), 155.

10. R. Fletcher and C. M. Reeves, Function Minimization by Conjugate Gradients, *Computer J.*, *7*(1964), 151.

11. J. Kowalik, Nonlinear Programming Procedures and Design Optimization, *Acta Polytech. Scandinav.*, *Math. Comp. Mach. Ser.*, No. 13 (1966).

12. J. A. Nelder and R. Mead, A Simplex Method for Function Minimization, *Computer J.*, *7*(1965), 308.

13. R. Fletcher, Certification of Algorithm 251 [E4], *Comm. ACM*, *9*(1966), 686.

14. For applications of the SUMT transformation see reference 11 and also:

G. P. McCormick, W. C. Nylander, and A. V. Fiacco, *Computer Program Implementing the Sequential Unconstrained Minimization Technique for Nonlinear Programming*, Research Analysis Corporation, Tech. Paper RAC-TP-151, McLean, Va., April, 1965.

A. Charnes, The SUMT Method for Convex Programming: Some Discussion and Experience, in *Recent Advances in Optimization Techniques*, A. Lavi and T. P. Vogl, eds. John Wiley & Sons, New York, 1996, p. 215.

B. C. Rush, J. Brodien, and G. P. McCormick, A Nonlinear Programming Model for Launch Vehicle Design and Costing, *Operations Res.*, *15* (1967), 185.

A. V. Fiacco, G. P. McCormick, and W. C. Nylander, Nonlinear Programming, Duality and Shadow Pricing by Sequential Unconstrained Optimization, paper presented to First World Econometric Congress, Rome, September 14, 1967.

D. Kavlie, J. Kowalik, S. Lund, and J. Moe, Design Optimization Using a General Nonlinear Programming Method, *European Shipbuilding*, *4*(1966), 1.

J. Moe and S. Lund, Cost and Weight Minimization of Structures with Special Emphasis on Longitudinal Strength Members of Tankers, paper presented to Summer Meeting of the Royal Institution of Naval Architects and Koninklijk Instituut van Ingenieurs, Holland, 1967.

Appendix I

SUMMARY OF MATRIX FORMULAS AND NOTATION

Here we summarize for completeness the notation and certain of the matrix formulas used in our development of optimization techniques. Further information can be found (for example) in Householder [1] and Fox [2].

We write $\mathbf{v}$ for the column vector with n components $v_1, v_2, \ldots, v_n$ where n will frequently not be specified explicitly. In these cases its value should follow either from the sense of the text, or from the requirement that matrix by vector multiplication be conformable. The row vector formed by transposing $\mathbf{v}$ is denoted by $\mathbf{v}^T$.

It is necessary to have a measure of the magnitude of $\mathbf{v}$, and we use here the *Euclidean norm* of $\mathbf{v}$ defined by

$$\|\mathbf{v}\| = \left\{ \sum_{i=1}^{n} v_i^2 \right\}^{1/2}. \tag{A1.1}$$

It is readily verified that $\|\mathbf{v}\|$ satisfies

$P1$: $\|\mathbf{v}\| = 0$ only if $v_i = 0$, $i = 1, 2, \ldots, n$,

$P2$: $\|\mathbf{u} + \mathbf{v}\| \leqslant \|\mathbf{u}\| + \|\mathbf{v}\|$,

$P3$: $\|\alpha \, \mathbf{v}\| = |\alpha| \, \|\mathbf{v}\|$.

Vectors of unit norm are called unit vectors. In particular the unit vectors parallel to the coordinate axes are denoted by $\mathbf{e}_i$, $i = 1, \ldots, n$. They have components defined by

$$(\mathbf{e}_i)_j = 0, \qquad j \neq i, \qquad (\mathbf{e}_i)_i = 1. \tag{A1.2}$$

We denote by A the matrix of the two-dimensional array of numbers A_{ij}, $i = 1, 2, \ldots, n$, $j = 1, 2, \ldots, m$. On occasion it is convenient to treat the individual rows and columns of A as vectors, and for this purpose we use the notation

$$\kappa_j(A) = j\text{-th column of } A = A_{ij}, \qquad i = 1, 2, \ldots, n,$$

$$\rho_s(A) = s\text{-th row of } A = A_{sj}, \qquad j = 1, 2, \ldots, m.$$

The matrix formed by transposing the rows and columns of A is denoted by A^T. If A is square (so that $m = n$) then the columns span an n-dimensional volume. The numerical value of this is called the determinant and is denoted by det (A).

If A is symmetric $(A = A^T)$ then we say that A is *positive-definite* if for any vector $\mathbf{x}$

    (i) $\mathbf{x}^T A \mathbf{x} \geqslant 0$, and

    (ii) $\mathbf{x}^T A \mathbf{x} = 0 \Rightarrow \mathbf{x} = 0$.

If condition (ii) does not hold then A is *positive-semidefinite* and det $(A) = 0$.

By putting $\mathbf{x} = \mathbf{e}_i$ it follows that

$$A_{ii} = \mathbf{e}_i^T A \mathbf{e}_i > 0, \tag{A1.3}$$

and by putting $\mathbf{x} = \mathbf{e}_i \pm \mathbf{e}_j$ that

$$A_{ii} + A_{jj} - 2A_{ij} > 0 \tag{A1.4}$$

and

$$A_{ii} + A_{jj} + 2A_{ij} > 0. \tag{A1.5}$$

Thus the diagonal elements of a positive-definite matrix are positive, and the element of largest modulus in the matrix lies on the diagonal.

The matrix I defined by

$$\kappa_j(I) = \mathbf{e}_j, \qquad j = 1, 2, \ldots, n,$$

is called the unit matrix. If A is square and $A^T A = I$ then A is said to be *orthogonal*. In this case det $(A) = \pm 1$.

It is necessary to have a measure for the magnitude of a square matrix, and it is convenient to relate this to the vector norm we use. For this reason we define $\|A\|$ by

$$\|A\| = \sup_{\mathbf{x} \neq 0} \frac{\|A\mathbf{x}\|}{\|\mathbf{x}\|} \tag{A1.6}$$

We note that

$$\|A\mathbf{x}\| \leqslant \|A\| \, \|\mathbf{x}\| \tag{A1.7}$$

and

$$\|AB\| \leqslant \|A\| \, \|B\|. \tag{A1.8}$$

On occasion we find it convenient to assemble a matrix from submatrices or to decompose it into submatrices. For example let

$$\mathbf{x} = \left| \begin{matrix} \mathbf{u} \\ \mathbf{v} \end{matrix} \right|, \qquad C = |A \mid B|,$$

then

$$C\mathbf{x} = A\mathbf{u} + B\mathbf{v}, \tag{A1.9}$$

assuming that the relevant matrices and vectors are conformable.

Most techniques for the solution of systems of linear equations which give an answer in a finite number of operations rely on a factorization of the matrix into the form

$$A = BU,$$

where the inverse of B is easily found and where U is upper triangular (that is $U_{ij} = 0$ if $i < j$). The solution of the system

$$A\mathbf{x} = \mathbf{b} \tag{A1.10}$$

is then found by solving in sequence

$$B\mathbf{y} = \mathbf{b}, \tag{A1.11}$$

$$U\mathbf{x} = \mathbf{y}. \tag{A1.12}$$

In the second step the solution is found by the back substitution process

$$x_n = y_n/U_{nn},$$

$$x_{n-1} = (y_{n-1} - U_{(n-1)n}x_n)/U_{(n-1)(n-1)},$$

and so on. If B is lower triangular (so that $B_{ij} = 0$ if $i > j$), the solution of the first stage is found in similar fashion by a forward substitution in which y_1 is calculated followed by y_2 and so on.

In generating techniques of this kind the class of elementary matrices has proved particularly valuable. These matrices have the form

$$E(\mathbf{u}, \mathbf{v}, \sigma) = I - \sigma\mathbf{u}\mathbf{v}^T, \tag{A1.13}$$

and have the property that

$$E(\mathbf{u}, \mathbf{v}, \sigma)E(\mathbf{u}, \mathbf{v}, \tau) = I - (\sigma + \tau - \sigma\tau\mathbf{v}^T\mathbf{u})\mathbf{u}\mathbf{v}^T$$
$$= E(\mathbf{u}, \mathbf{v}, \sigma + \tau - \sigma\tau\mathbf{v}^T\mathbf{u}). \tag{A1.14}$$

In particular if

$$\tau = \frac{-\sigma}{1 - \sigma\mathbf{v}^T\mathbf{u}} \tag{A1.15}$$

then the second matrix is the inverse of the first. It is this property of having a readily computable inverse that makes elementary matrices convenient in practice.

REFERENCES

1. A. S. Householder, *The Theory of Matrices in Numerical Analysis*, Blaisdell Publishing Company, Waltham, Mass., 1964.
2. L. Fox, *An Introduction to Numerical Linear Algebra*, Oxford University Press, London and New York, 1964.

Appendix 2

SOME RESULTS AND DEFINITIONS CONCERNING CONVEXITY

On a number of occasions (in Chapters 3 and 5 in particular) we have need of certain properties of convex sets and functions, and these are summarized here.

DEFINITION 1.   *Convex set of points:* the set S convex if for every two points $\mathbf{x}^{(1)}$, $\mathbf{x}^{(2)}$ in the set the line segment joining these points is also in the set, i.e.,

$$\text{if } \mathbf{x}^{(1)}, \mathbf{x}^{(2)} \in S, \qquad \text{then } \mathbf{x} \in S,$$

where

$$\mathbf{x} = \lambda\mathbf{x}^{(2)} + (1 - \lambda)\mathbf{x}^{(1)}, \qquad 0 \leqslant \lambda \leqslant 1. \tag{A2.1}$$

By assumption the set containing only one point is convex.

THEOREM 1.   *An intersection of any number of convex sets is also convex.*

PROOF.   Let R^i, $i = 1, 2, \ldots, k$, be convex and let $\mathbf{x}^{(1)}$ and $\mathbf{x}^{(2)}$ belong to $R = \cap_{i=1}^{k} R_i$. Any $(1 - \lambda)\mathbf{x}^{(1)} + \lambda\mathbf{x}^{(2)} = \mathbf{x} \in R_i$ for all i and $0 \leqslant \lambda \leqslant 1$. Hence $\mathbf{x} \in R$ and the theorem is shown.

We assume that $f(\mathbf{x})$ is differentiable with continuous partial derivatives, i.e., the components of

$$\nabla f(x) = \left[\frac{\partial f(\mathbf{x})}{\partial x_1}, \frac{\partial f(\mathbf{x})}{\partial x_2}, \ldots, \frac{\partial f(\mathbf{x})}{\partial x_n} \right] \tag{A2.2}$$

are continuous functions of $\mathbf{x}$.

DEFINITION 2.   *Convex function:* The function $f(\mathbf{x})$ is convex if a linear interpolation between values of the function never underestimates the real value of the function taken at the interpolated point, i.e.,

$$f[(1 - \lambda)\mathbf{x}^{(2)} + \lambda\mathbf{x}^{(1)}] \leqslant (1 - \lambda)f(\mathbf{x}^{(2)}) + \lambda f(\mathbf{x}^{(1)}), \tag{A2.3}$$

where $0 \leqslant \lambda \leqslant 1$.

DEFINITION 3.   *Concave function:* The function $f(\mathbf{x})$ is concave if $-f(\mathbf{x})$ is convex.

DEFINITION 4. *Strict convexity (concavity):* The function $f(\mathbf{x})$ is strictly convex (concave) if strict inequality holds in (A2.3) for any $\mathbf{x}^{(1)} \neq \mathbf{x}^{(2)}$.

THEOREM 2. *The following definitions of convexity are equivalent to Definition 2:*

(a) For any two points $\mathbf{x}^1$, $\mathbf{x}^2$ we have

$$f(\mathbf{x}^{(2)}) \geqslant f(\mathbf{x}^{(1)}) + \nabla^T f(\mathbf{x}^{(1)})(\mathbf{x}^{(2)} - \mathbf{x}^{(1)}). \tag{A2.4}$$

PROOF. Let $f(\mathbf{x})$ be convex. Then

$$f[(1 - \lambda)\mathbf{x}^{(1)} + \lambda\mathbf{x}^{(2)}] \leqslant (1 - \lambda)f(\mathbf{x}^{(1)}) + \lambda f(\mathbf{x}^{(2)}), \tag{A2.5}$$

which gives

$$f[\mathbf{x}^{(1)} + \lambda(\mathbf{x}^{(2)} - \mathbf{x}^{(1)})] \leqslant f(\mathbf{x}^{(1)}) + \lambda[f(\mathbf{x}^{(2)}) - f(\mathbf{x}^{(1)})], \tag{A2.6}$$

and further

$$f(\mathbf{x}^2) - f(\mathbf{x}^1) \geqslant \frac{f[\mathbf{x}^{(1)} + \lambda(\mathbf{x}^{(2)} - \mathbf{x}^{(1)})] - f(\mathbf{x}^{(1)})}{\lambda}. \tag{A2.7}$$

Relationship (A2.4) follows from (A2.7) for $\lambda \to 0$.

(b) If we assume that $f(\mathbf{x})$ is twice continuously differentiable over a convex set S, then $f(\mathbf{x})$ is convex in S if and only if the Hessian matrix defined by

$$H_{ij}(\mathbf{x}) = \frac{\partial^2 f(\mathbf{x})}{\partial x_i \, \partial x_j} \tag{A2.8}$$

is positive-semidefinite for any $\mathbf{x}$ in S.

PROOF. From the Taylor theorem we have that for any two points $\mathbf{x}^{(1)}$, $\mathbf{x}^{(2)} \in S$ there exists $0 \leqslant \theta \leqslant 1$, such that

$$f(\mathbf{x}^{(2)}) = f(\mathbf{x}^{(1)}) + (\nabla f(\mathbf{x}^{(1)}))^T (\mathbf{x}^{(2)} - \mathbf{x}^{(1)})$$
$$+ \tfrac{1}{2}(\mathbf{x}^{(2)} - \mathbf{x}^{(1)})^T H[\theta\mathbf{x}^{(2)} + (1 - \theta)\mathbf{x}^{(1)}](\mathbf{x}^{(2)} - \mathbf{x}^{(1)}). \tag{A2.9}$$

But

$$f(\mathbf{x}^{(2)}) - f(\mathbf{x}^{(1)}) - (\nabla f(\mathbf{x}^{(1)}))^T (\mathbf{x}^{(2)} - \mathbf{x}^{(1)})$$
$$= \tfrac{1}{2}(\mathbf{x}^{(2)} - \mathbf{x}^{(1)})H(\theta, \mathbf{x}^{(1)}, \mathbf{x}^{(2)})(\mathbf{x}^{(2)} - \mathbf{x}^{(1)}) \geqslant 0. \tag{A2.10}$$

Using the fact that $H(\mathbf{x})$ is a continuous function of $\mathbf{x}$, the necessity of the positive-semidefiniteness of this matrix follows. Clearly, $f(\mathbf{x})$ is strictly convex under the condition that $H(\mathbf{x})$ is positive-definite.

THEOREM 3. *A positive combination of convex functions is again convex and strictly convex if at least one of the functions is strictly convex.*

The proof is immediate.

THEOREM 4. *The set of points S which satisfies the constraint $g(\mathbf{x}) \geqslant 0$, where $g(\mathbf{x})$ is a concave function, is a convex set.*

PROOF. Let $\mathbf{x}^{(1)}$ and $\mathbf{x}^{(2)}$ be two points of S, i.e.,

$$g(\mathbf{x}^{(1)}) \geqslant 0, \qquad g(\mathbf{x}^{(2)}) \geqslant 0.$$

We have for $0 \leqslant \lambda \leqslant 1$

$$g[(1 - \lambda)\mathbf{x}^{(2)} + \lambda\mathbf{x}^{(1)}] \geqslant g(\mathbf{x}^{(1)}) + (1 - \lambda)g(\mathbf{x}^{(2)}) \geqslant 0,$$

which shows that any $\mathbf{x} = \lambda\mathbf{x}^{(1)} + (1 - \lambda)\mathbf{x}^{(2)}$ belongs to S.

It follows from Theorems 4 and 1 that the set $R^0 = \{\mathbf{x} \mid g_i(\mathbf{x}) \geqslant 0, i = 1, 2, \ldots, I\}$ is a convex set if all $g_i(\mathbf{x})$ are concave.

THEOREM 5. *If the function $g(\mathbf{x})$ is concave in R^0 then $g^{-1}(\mathbf{x})$ is convex in this region.*

PROOF. It is sufficient to show that the Hessian of $g^{-1}(\mathbf{x})$ is positive-semidefinite for any $\mathbf{x} \in R^0$ where $g(\mathbf{x}) \geqslant 0$.

We have

$$H_{ij}(\mathbf{x}) = \frac{\partial^2(g^{-1}(\mathbf{x}))}{\partial x_i \, \partial x_j} = g^{-2}(\mathbf{x})\frac{\partial^2(-g(\mathbf{x}))}{\partial x_i \, \partial x_j} + 2g^{-3}(\mathbf{x})\frac{\partial g(\mathbf{x})}{\partial x_i}\frac{\partial g(\mathbf{x})}{\partial x_j} \quad \text{(A2.11)}$$

and

$$H(\mathbf{x}) = g^{-2}(\mathbf{x})H_1(\mathbf{x}) + 2g^{-3}(\mathbf{x})H_2(\mathbf{x}). \quad \text{(A2.12)}$$

Clearly $H_1(\mathbf{x})$ is positive-semidefinite, because $-g(\mathbf{x})$ is convex. The second matrix is

$$H_2(\mathbf{x}) = (\nabla g(\mathbf{x}))(\nabla g(\mathbf{x}))^T, \quad \text{(A2.13)}$$

and for an arbitrary $\mathbf{V}$ we have

$$\mathbf{V}^T H_2(\mathbf{x})\mathbf{V} = Z^2 \geqslant 0, \quad \text{(A2.14)}$$

where

$$Z = \mathbf{V}^T \nabla g(\mathbf{x}). \quad \text{(A2.15)}$$

Since a positive sum of positive-semidefinite matrices is positive-semidefinite, the theorem is proved.

Note that from Theorems 3 and 5 it follows that, if $f(\mathbf{x})$, $-g_1(\mathbf{x})$, $-g_2(\mathbf{x})$, \ldots, $-g_I(\mathbf{x})$ are convex, then for any $r > 0$ the function

$$T(\mathbf{x}, r) = f(\mathbf{x}) + r \sum_i g_i^{-1}(\mathbf{x}) \quad \text{(A2.16)}$$

is convex.

DEFINITION 5.   *Global minimum:* A point $\bar{\mathbf{x}}$ of a domain R on which a function $f(\mathbf{x})$ is defined is a global minimum in R if for any $\mathbf{x} \in R$ the following inequality holds

$$f(\bar{\mathbf{x}}) \leqslant f(\mathbf{x}), \qquad \mathbf{x} \in R. \tag{A2.17}$$

DEFINITION 6.   *Local minimum:* The function $f(\mathbf{x})$ has a local minimum at $\bar{\mathbf{x}}$ if there exists an $\varepsilon > 0$ such that for all $\mathbf{x}$ in $S = \{\mathbf{x} \mid 0 < \|\mathbf{x} - \bar{\mathbf{x}}\| < \varepsilon\}$ we have

$$f(\bar{\mathbf{x}}) \leqslant f(\mathbf{x}), \qquad \mathbf{x} \in S. \tag{A2.18}$$

THEOREM 6.   *Any local minimum of a convex function $f(\mathbf{x})$ attained on a convex set S is a global minimum.*

PROOF.   Suppose that $\mathbf{x}^{(1)}$, $\mathbf{x}^{(2)} \in S$ which is a convex set and both are local minima.

Let us assume that $f(\mathbf{x}^{(2)}) < f(\mathbf{x}^{(1)})$; then it follows from the convexity of $f(\mathbf{x})$ that

$$(\nabla f(\mathbf{x}^{(1)}))^T(\mathbf{x}^{(2)} - \mathbf{x}^{(1)}) \leqslant f(\mathbf{x}^{(2)}) - f(\mathbf{x}^{(1)}) \tag{A2.19}$$

and

$$(\nabla f(\mathbf{x}^{(1)}))^T(\mathbf{x}^{(2)} - \mathbf{x}^{(1)}) \leqslant 0. \tag{A2.20}$$

We see that $f(\mathbf{x})$ would decrease, if we move from $\mathbf{x}^{(1)}$ in the direction $\mathbf{s} = \mathbf{x}^{(2)} - \mathbf{x}^{(1)}$. Since the segment $\mathbf{x} = (1 - \lambda)\mathbf{x}^{(1)} + \lambda_2\mathbf{x}^{(2)}$, $0 \leqslant \lambda \leqslant 1$, belongs to S, it follows that $\mathbf{x}^{(1)}$ could not be a minimum.

THEOREM 7.   *The strictly convex function $f(\mathbf{x})$ has only one minimum on a convex set S or is unbounded below.*

PROOF.   Let $\mathbf{x}^{(1)}$, $\mathbf{x}^{(2)} \in S$ be two minima and $f(\mathbf{x}^1) = f(\mathbf{x}^2)$.

From the strict convexity assumption it follows that any point $\mathbf{x} = (1 - \lambda)\mathbf{x}^{(1)} + \lambda\mathbf{x}^{(2)}$, $0 \leqslant \lambda \leqslant 1$ will have a lower function value so that there cannot be two minima.

NOTES ON RECENT DEVELOPMENTS

1. Zangwill (1) has drawn attention to a disadvantage of the simple form of Powell's conjugate direction method without derivatives and has given an example to show that the directions generated need not be linearly independent. This observation does not apply to the modified Powell algorithm. Zangwill suggests an algorithm which is similar to our method using restarting so that the comments in Chapter 6 are likely to be relevant.

2. In Chapter 4 we remarked that the quadratically convergent variant of the Gauss algorithm had attracted little attention. However, Goldstein and Price (2) have given a method in which the second derivative terms are approximated by finite differences. Some results concerning the convergence of this algorithm are proved.

3. Stewart (3) has given a modification of the Davidon method in which the derivatives are calculated by difference methods. Particular attention is paid to the competing requirements of high accuracy and low cancellation errors. Stewart claims that the method is superior to the Powell conjugate direction method without derivatives.

4. We have referred to the pleasing development of our subject to the stage in which certain aspects at least permit a unifying treatment. For the case of the quasi-Newton methods (see, for example, our treatment of the Davidon and secant algorithms), a comprehensive derivation has been given by Zeleznik (4).

   Davidon (5) has produced an interesting new method in this class in which the correction matrix has rank 1 (not 2 as in conjugate direction method) and which finds the minimum of a quadratic form in $n + 2$ steps.

   A further feature of this method is that minimization along a line is not necessary as a full Newton step is used if this is successful in reducing the objective function, and otherwise the direction of search is changed in a prescribed manner. Apparently this algorithm currently lacks a definitive implementation.

5. The current interest in solving constrained problems by a sequence of suitably transformed unconstrained problems has led to further consideration being given to cases in which the objective function is not adequately approximated by a positive definite quadratic form in a neighborhood of the minimum [Murray (6)]. Powell (7) has considered widening the range of convergence of Newton's method by using steepest-descent steps when the Jacobian matrix is ill-conditioned. He has provided an automatic procedure for choosing the search direction.

6. Morrison (8) has given an alternative treatment of problems with equality constraints by introducing the transformed problem to minimize

$$F(\mathbf{x}, M) = (f(\mathbf{x}) - M)^2 + \sum_{i=1}^{p} g_i^2(\mathbf{x}),$$

where $f(\mathbf{x})$ is the objective function and $g_i(\mathbf{x}) = 0$, $i = 1, 2, \ldots, p$ are constraints. Let $\bar{f}$ and $\bar{\mathbf{x}}$ be the optimum solution of the constrained problem. Morrison gives the following algorithm (assuming $M_1 < \bar{f}$)

(i) Find $\mathbf{x}_k$, the unconstrained minimum of $F(\mathbf{x}, M_k)$,
(ii) Set $M_{k+1} = M_k + (F(\mathbf{x}_k, M_k))^{1/2}$,
(iii) If $M_{k+1} = M_k$, stop. Otherwise repeat from (i) with $k: = k + 1$.

He shows that M_k increases monotonically to $\bar{f}$. We remark that this algorithm complements the algorithm by Schmit and Fox (Chapter 5) in which pessimistic estimates of $\bar{f}$ are required.

Also Murray has pointed out to us that the Hessian of $F(\mathbf{x}, M)$ approaches singularity as $M_k \to \bar{f}$, so that conjugate direction methods would tend to lose efficiency as the minimum is found (see our remarks on the SUMT transformation in Chapter 6). However, note that M_k is already a good approximation of $\bar{f}$ when this trouble occurs.

7. Powell (9) criticizes the SUMT transformation (in particular in the case of equality constraints) because of the tendency of the quantities computed to vary substantially in magnitude leading to heavy cancellation as the calculation proceeds. He suggests a transformation which is somewhat similar but in which further parameters are provided. This transformation is

$$\Phi(\mathbf{x}, \boldsymbol{\sigma}, \boldsymbol{\theta}) = f(\mathbf{x}) + \sum_{i=1}^{p} \sigma_i(g_i(\mathbf{x}) + \theta_i)^2,$$

where $f(\mathbf{x})$ is the objective function and the equality constraints are $g_i(\mathbf{x}) = 0$, $i = 1, 2, \ldots, p$.

The method is based on the observation that if $\bar{\mathbf{x}}(\sigma, \theta)$ minimizes Φ, then the same point minimizes $f(\mathbf{x})$ subject to constraints $g_i(\mathbf{x}) = g_i(\bar{\mathbf{x}})$. Thus, the problem reduces to finding values of σ, θ such that $g_i(\bar{\mathbf{x}}(\sigma, \theta)) = 0$, $i = 1, 2, \ldots, p$.

Powell is hopeful that this method can be extended to apply also to the case of inequality constraints.

A transformation similar to the SUMT method has been suggested by Murray (10). His approach differs from the already described in that he uses quadratic programming to generate the directions for a descent step at each stage of the iteration.

REFERENCES

1. W. I. Zangwill, Minimizing a Function Without Calculating Derivatives, Computer J., *10*(1967), 293.
2. A. A. Goldstein, and J. F. Price, An Effective Algorithm for Minimization, Numerische Mathematik, *10*(1967), 184.
3. G. W. Stewart, A Modification of Davidon's Minimization Method to Accept Difference Approximations of Derivatives, J.A.C.M., *14*(1967), 72.
4. F. J. Zeleznik, Quasi-Newton Methods for Nonlinear Equations, J.A.C.M., *15*(1968), 265.
5. W. C. Davidon, Variance Algorithm for Minimization, Computer J., *10*(1968), 406.
6. W. Murray, Ill-conditioning in Barrier and Penalty Functions Arising in Constrained Nonlinear Programming, to appear in Proceedings of the Sixth International Symposium on Mathematical Programming, Princeton University, August 13–20, 1967.
7. M. J. D. Powell, Private communication.
8. D. D. Morrison, Optimization by Least Squares, SIAM J., Numer. Anal., *5*(1968), 83.
9. M. J. D. Powell, A Method for Non-linear Constraints in Minimization Problems, T. P. 310, Atomic Energy Research Establishment, Harwell, England, 1967.
10. W. Murray, An Algorithm for Constrained Minimization, National Physical Laboratory, Teddington, England, unpublished paper, 1968.

AUTHOR INDEX

Page numbers set in *italics* denote the pages on
which the complete literature reference are given.

SUBJECT INDEX